Springer Theses

Recognizing Outstanding Ph.D. Research

Aims and Scope

The series "Springer Theses" brings together a selection of the very best Ph.D. theses from around the world and across the physical sciences. Nominated and endorsed by two recognized specialists, each published volume has been selected for its scientific excellence and the high impact of its contents for the pertinent field of research. For greater accessibility to non-specialists, the published versions include an extended introduction, as well as a foreword by the student's supervisor explaining the special relevance of the work for the field. As a whole, the series will provide a valuable resource both for newcomers to the research fields described, and for other scientists seeking detailed background information on special questions. Finally, it provides an accredited documentation of the valuable contributions made by today's younger generation of scientists.

Theses may be nominated for publication in this series by heads of department at internationally leading universities or institutes and should fulfill all of the following criteria

- They must be written in good English.
- The topic should fall within the confines of Chemistry, Physics, Earth Sciences, Engineering and related interdisciplinary fields such as Materials, Nanoscience, Chemical Engineering, Complex Systems and Biophysics.
- The work reported in the thesis must represent a significant scientific advance.
- If the thesis includes previously published material, permission to reproduce this must be gained from the respective copyright holder (a maximum 30% of the thesis should be a verbatim reproduction from the author's previous publications).
- They must have been examined and passed during the 12 months prior to nomination.
- Each thesis should include a foreword by the supervisor outlining the significance of its content.
- The theses should have a clearly defined structure including an introduction accessible to new PhD students and scientists not expert in the relevant field.

Indexed by zbMATH.

More information about this series at http://www.springer.com/series/8790

Yuki Sato

Quantum Oscillations and Charge-Neutral Fermions in Topological Kondo Insulator YbB_{12}

Doctoral Thesis accepted by
Kyoto University, Kyoto, Japan

Springer

Author
Dr. Yuki Sato
Center for Emergent Matter Science (CEMS)
RIKEN
Wako, Saitama, Japan

Supervisor
Prof. Yuji Matsuda
Department of Physics
Kyoto University
Kyoto, Japan

ISSN 2190-5053 ISSN 2190-5061 (electronic)
Springer Theses
ISBN 978-981-16-5676-7 ISBN 978-981-16-5677-4 (eBook)
https://doi.org/10.1007/978-981-16-5677-4

This Springer imprint is published by the registered company Springer Nature Singapore Pte Ltd.
The registered company address is: 152 Beach Road, #21-01/04 Gateway East, Singapore 189721, Singapore

Supervisor's Foreword

This Ph.D. thesis describes an investigation conducted by Dr. Yuki Sato during his doctoral program at the Graduate School of Science, Kyoto University. In his graduate study, he achieved considerable results in the research of the Kondo insulator, which belongs to the group of strongly correlated electron systems. In particular, he made several important discoveries on the exotic physical properties of YbB_{12}, a topological Kondo insulator. As the supervisor of the doctoral course, here I introduce two most remarkable findings, namely quantum oscillations and neutral fermion excitations in the insulator.

Fermi surfaces in metallic systems can be detected by observing characteristic quantum oscillations of magnetization and electrical resistivity in a strong magnetic field—they are consequences of the quantization of the orbital motion of conduction electrons. Such oscillations, in particular of the resistivity, are absent in insulators, because of the absence of Fermi surfaces. Dr. Sato discovered a profoundly controversial behavior in a Kondo insulator YbB_{12}. This material is unmistakably an insulator with an electric resistivity magnitude far beyond that of metals. However, he discovered that its resistivity exhibits quantum oscillations. Analysis of the oscillations reveals the three-dimensional nature of the Fermi surface, demonstrating that the signal arises from the bulk rather than the surface. Although the oscillations of the resistivity and magnetization are observed in an insulator, their amplitude follows the conventional Fermi liquid theory of metals. The large effective masses determined by the oscillations point to the presence of Fermi surfaces consisting of strongly correlated electrons.

Dr. Sato also discovered the gapless itinerant excitations of charge-neutral fermions in the ground state of YbB_{12}. Charge carriers in metals convey both electronic and heat currents due to the presence of the Fermi surface. In almost all metallic systems, the ratio of electric and thermal conductivities is constant, which is known as Wiedemann–Franz law. By performing the very low-temperature heat-transport measurements on YbB_{12} at low magnetic fields, Dr. Sato revealed a striking violation of this law, demonstrating that this compound is an electrical insulator but a "thermal metal." Moreover, the more insulating crystal exhibits the more metallic thermal properties, i.e., the opposite of conventional metal physics. In addition, despite the charge neutrality, the neutral fermions can couple to a magnetic field.

These unexpected results demonstrate novel aspects of strongly correlated insulators. I hope that there will be many readers who will gain a broader perspective of strongly correlated and topological systems in these fascinating systems as a result of Dr. Sato's efforts.

Kyoto, Japan
July 2021

Prof. Yuji Matsuda

Publication List

1. Y. Sato, S. Kasahara, H. Murayama, Y. Kasahara, E.-G. Moon, T. Nishizaki, T. Loew, J. Porras, B. Keimer, T. Shibauchi, and Y. Matsuda, "Thermodynamic evidence for a nematic phase transition at the onset of the pseudogap in $YBa_2Cu_3O_y$," Nat. Phys. **13**, 1074-1078 (2017).
2. Y. Sato, S. Kasahara, T. Taniguchi, X. Xing, Y. Kasahara, Y. Tokiwa, Y. Yamakawa, H. Kontani, T. Shibauchi, and Y. Matsuda, "Abrupt change of the superconducting gap structure at the nematic critical point in $FeSe_{1-x}S_x$," Proc. Natl. Acad. Sci. USA **115**, 1227-1231 (2018).
3. Z. Xiang, Y. Kasahara, T. Asaba, B. Lawson, C. Tinsman, Lu Chen, K. Sugimoto, S. Kawaguchi, Y. Sato, G. Li, S. Yao, Y. L. Chen, F. Iga, J. Singleton, Y. Matsuda, and Lu Li, "Quantum oscillations of electrical resistivity in an insulator," Science **69**, 65-69 (2018).
4. H. Murayama, Y. Sato, R. Kurihara, S. Kasahara, Y. Mizukami, Y. Kasahara, H. Uchiyama, A. Yamamoto, E. G. Moon, J. Cai, J. Freyermuth, M. Greven, T. Shibauchi, and Y. Matsuda, "Diagonal nematicity in the pseudogap phase of $HgBa_2CuO_{4+\delta}$," Nat. Commun. **10**, 3282 (2019).
5. H. Murayama, Y. Sato, X. Z. Xing, T. Taniguchi, S. Kasahara, Y. Kasahara, M. Yoshida, Y. Iwasa, and Y. Matsuda, "Effect of quenched disorder on the quantum spin liquid state of the triangular-lattice antiferromagnet 1T-TaS_2," Phys. Rev. Res. **2**, 013099 (2020).
6. S. Kasahara, Y. Sato, S. Licciardello, M. Culo, S. Arsenijevic, T. Ottenbros, T. Tominaga, J. Böker, I. Eremin, T. Shibauchi, J. Wosnitza, N. E. Hussey, and Y. Matsuda, "Evidence for an Fulde-Ferrell-Larkin-Ovchinnikov State with Segmented Vortices in the BCS-BEC-Crossover Superconductor FeSe," Phys. Rev. Lett. **124**, 107001 (2020).
7. M. Yamashita, Y. Sato, T. Tominaga, Y. Kasahara, S. Kasahara, H. Cui, R. Kato, T. Shibauchi, and Y. Matsuda, "Presence and absence of itinerant gapless excitations in the quantum spin liquid candidate $EtMe_2Sb[Pd(dmit)_2]_2$," Phys. Rev. B, **101**, 140407(R) (2020).
8. W. K. Huang, S. Hosoi, M. Culo, S. Kasahara, Y. Sato, K. Matsuura, Y. Mizukami, M. Berben, N. E. Hussey, H. Kontani, T. Shibauchi, and Y. Matsuda,

"Non-Fermi liquid transport in the vicinity of the nematic quantum critical point of superconducting $FeSe_{1-x}S_x$," Phys. Rev. Res. **2**, 033367 (2020).
9. H. Murayama, K. Ishida, R. Kurihara, T. Ono, Y. Sato, Y. Kasahara, H. Watanabe, Y. Yanase, G. Cao, Y. Mizukami, T. Shibauchi, Y. Matsuda, and S. Kasahara, "Bond directional anapole order in a spin-orbit coupling Mott insulator $Sr_2\,(Ir_{1-x}Rh_x)O_4$," Phys. Rev. X **11**, 011021 (2021).
10. Z. Xiang, Lu Chen, K.-W. Chen, C. Tinsman, Y. Sato, T. Asaba, H. Lu, Y. Kasahara, M. Jaime, F. Balakirev, F. Iga, Y. Matsuda, J. Singleton, and Lu Li, "Unusual high-field metal in a Kondo insulator," Nat. Phys. **17**, 788–793 (2021).
11. Y. Sato, S. Suetsugu, T. Tominaga, Y. Kasahara, S. Kasahara, T. Kobayashi, S. Kitagawa, K. Ishida, R. Peters, T. Shibauchi, A. H. Nevidomskyy, L. Qian, J. M. Moya, E. Morosan, and Y. Matsuda, "Charge neutral fermions and magnetic field driven instability in insulating $YbIr_3Si_7$," Preprint at arXiv:2103.13718 (2021).

Parts of this thesis have been published in the following journal articles:

1. Y. Sato, Z. Xiang, Y. Kasahara, S. Kasahara, Lu Chen, C. Tinsman, F. Iga, J. Singleton, N. L. Nair, N. Maksimovic, J. G. Analytis, Lu Li, and Y. Matsuda, "Topological surface conduction in Kondo insulator YbB_{12}," J. Phys. D: Appl. Phys. **54**, 404002 (2021).
2. Y. Sato, Z. Xiang, Y. Kasahara, T. Taniguchi, S. Kasahara, Lu Chen, T. Asaba, C. Tinsman, H. Murayama, O. Tanaka, Y. Mizukami, T. Shibauchi, F. Iga, J. Singleton, Lu Li, and Y. Matsuda, "Unconventional thermal metallic state of charge-neutral fermions in an insulator," Nat. Phys. **15**, 954 (2019).

Preface

Strong electronic correlation in many-body systems often results in a wide variety of ground states, leading to phenomena such as heavy Fermi liquids, unconventional superconductivity, and strongly correlated insulators. Understanding those exotic electronic phases and excitations (i.e., emergent quasiparticles) from these ground states is the most fundamental and important research problem in condensed matter physics. A Kondo insulator (KI) is a typical example of such strongly correlated insulators, where hybridization between conduction c-electrons and localized f-electrons opens up a charge gap across the Fermi level. KIs have recently attracted much interest because of the following important theoretical predictions and experimental findings:

(a) Strong electronic correlation can drive certain KIs into topological insulators.
(b) The KIs exhibit quantum oscillations (QOs) in high-magnetic fields.
(c) The KIs show gapless fermionic excitations.

In this monograph, we review exotic electronic properties found in topological Kondo insulator candidates SmB_6 and YbB_{12} and present our study which reveals the aforementioned remarks: YbB_{12} shows (a) topological surface conduction, (b) QOs both in magnetization and resistivity, and (c) gapless charge-neutral fermionic excitations. We stress that the observations of both (b) and (c), in general, indicate the presence of a Fermi surface, which has been considered a defining characteristic of metals. Our findings, therefore, demonstrate novel electronic properties in YbB_{12}: its bulk exhibits "metallic" behavior, although it is electrically insulating.

This book is structured as follows. In Chap. 1, we introduce the background physics of KIs and examine how the Kondo lattice systems can be classified into either KIs or heavy Fermi liquids (FLs). We also see basic properties of KIs and FLs. Subsequently, we review the electronic properties of the candidate materials SmB_6 and YbB_{12}, including recent progress in the search for topological surface states. In Chap. 2, we show textbook introduction of QOs and how Fermi surface can be experimentally determined. We then review previous reports and theoretical proposals on the observations of QOs in SmB_6 and examine how they can be interpreted. We also raise some issues regarding those findings. Next, in Chap. 3, we review the observation of QOs in YbB_{12} and examine how the observations are different from what is observed in SmB_6. We also show QOs observed in field-induced metallic phase in YbB_{12} and introduce two fluid model, which supports the scenario that the charge-neutral fermions are attributed to the QOs.

In Chap. 4, we show our systematic transport study of microstructures fabricated by focused ion beam method to investigate topological surface states in YbB_{12}. We prepared samples with carefully controlling their thickness and observed all sample shows saturation in the resistivity at low temperatures. Our analysis reveals that metallic surface conduction takes place on the insulating bulk. Moreover, the magnetoresistance of the microstructure exhibits weak-antilocalization effect, suggesting that the spin of the surface states is coupled to its momentum. Our study demonstrates that YbB_{12} host topological surface conduction, which is consistent with theoretical predictions. Despite of the presence pf topological surface states, the microstructure does not show Shubnikov–de Haas (SdH) oscillations in high-magnetic fields, while the single crystals show a distinct signal as discussed in Chap. 3. The result provides a supporting evidence that the SdH effect derives from the insulating bulk.

In Chap. 5, we present low-temperature heat-transport measurements to discuss low-energy excitations in the ground state of YbB_{12}, which is a major work conducted by the author. The presence of a Fermi surface is manifested in the linear temperature (T)-dependent terms in specific heat and thermal conductivity. At zero field, sizable linear T-dependent terms are clearly observed in the specific heat and thermal conductivity, indicating the presence of gapless fermionic excitations with an itinerant character. Remarkably, the observed linear T-dependent thermal conductivity leads to a spectacular violation of the Wiedemann–Franz law: the Lorenz ratio is 10^4–10^5 times larger than that expected in conventional metals, indicating that YbB_{12} is electrically insulating but thermally metallic. Interestingly, the neutral fermions become more mobile when the sample becomes more insulating, ruling out the possibility that minor metallic impurities contribute to the heat transport. Moreover, these fermions couple to magnetic fields, despite their charge neutrality. Our findings expose novel quasiparticles in this unconventional quantum state, which are potentially identical to what contributes to the QOs in the insulating phase.

Finally, we summarize and conclude the present work in Chap. 6.

Wako, Japan
July 2021

Yuki Sato

Acknowledgements

This book is based on my Ph.D. thesis and works in Department of Physics, Kyoto University. Firstly, I would like to express my greatest gratitude to my supervisor, Prof. Yuji Matsuda, for his continuous support throughout my doctoral course. Without his instructions and immense knowledge in physics, I certainly would not have been able to complete the study related to this monograph.

My sincere appreciation also goes to Prof. Takahito Terashima, Prof. Yuichi Kasahara, Prof. Shigeru Kasahara, Prof. Takasada Shibauchi, Prof. Yoshifumi Tokiwa, Prof. Lu Li, Prof. Eun-Gook Moon, Prof. Hiroshi Kontani, and Prof. Robert Peters for their experimental instructions and/or stimulating discussions related to my works in Kyoto. Further, I would like to appreciate Prof. John Singleton, Dr. Ziji Xiang, Dr. Lu Chen, Dr. Colin Tinsman, Dr. Ben Lawson, and Dr. Tomoya Asaba for discussions and especially their great efforts in the pulsed-field experiments at Los Alamos National Laboratory. Additionally, I would like to thank Prof. Fumitoshi Iga for providing extraordinarily clean single crystals of YbB_{12} and SmB_6. I am deeply indebted to following fellows for discussions and helpful experimental support: Hinako Murayama, Dr. Xiangzhuo Xing, Tomoya Taniguchi, Ryo Kurihara, Takahiro Tominaga, Dr. Huang Wenkai, Dr. Yuta Mizukami, Ohei Tanaka, Dr. Nityan Nair, and Nikola Maksimovic.

During my Ph.D. course, I was fortunate to have several chances to travel and conduct experiments abroad. I am deeply grateful to Prof. James Analytis for his kind support during my stay in Berkeley. I would also like to express my appreciation for Prof. Nigel Hussey and Dr. Matija Čulo for their kind support during intermittent experiments at High Field Magnet Laboratory in Nijimegen.

Many collaborators were involved in related unpublished work on another Kondo insulator, $YbIr_3Si_7$, which we have prepared but unfortunately could not be included in this book. I would like to thank Prof. Emilia Morosan and her workfellows in Rice University for providing us single crystals. I would also like to thank Prof. Andriy Nevidomskyy and Dr. Shota Suetsugu for fruitful discussions related to this work. Moreover, I thank Prof. Kenji Ishida and Dr. Shunsaku Kitagawa for conducting NMR experiments on $YbIr_3Si_7$.

I am grateful to my labmates not only for their cooperation for my research but also for sharing a tough but fruitful time in Kyoto: Dr. Takuya Yamashita, Dr. Tatsuya Watashige, Yusuke Shimoyama, Dr. Masahiro Naritsuka, Daiki Terazawa, Tomohiro Ishii, Takafumi Onishi, Yohei Torii, Hironori Hashimoto, Sohei Miyake, Dr. Masahiro Haze, Tomoka Suematsu, Ma Sixiao, Daiki Sano, Satoshi Nakamura, Dr. Lang Peng, Takahito Ii, Hiroki Suzuki, Toshifumi Takahashi, Andre de Oliveira Silva, Yuzuki Ukai, Hiroaki Ominato, Takahiro Ono, and Taichi Yokoi. I would also like to thank laboratory secretary Ayumi Ishikawa for handling paperwork related to my research.

I wish to acknowledge financial support from the Research Fellowships of the Japan Society for the Promotion of Science for Young Scientists during my entire doctoral course. I would also like to thank Editage (www.editage.com) for English language editing for my original thesis.

Lastly, I must thank my family for their warm encouragement and support.

July 2021 Yuki Sato

Contents

Acronyms

ARPES	Angle-resolved photoemission spectroscopy
BZ	Brillouin zone
CNF	Charge-neutral fermion
dHvA	de Haas–van Alphen
DOS	Density of states
FFT	Fast Fourier transformation
FIB	Focused ion beam
FL	Fermi liquid
HLN	Hikami–Larkin–Nagaoka
I-M	Insulator metal
IS	Inversion symmetry
KI	Kondo insulator
KM	Kondo metal
KW	Kadowaki–Wood
LDA	Local density approximation
LK	Lifshitz–Kosevich
LP	Low-pass
MR	Magnetoresistance
NMR	Nuclear magnetic resonance
PDO	Proximity-detector-oscillator
PES	Photoemission spectroscopy
QCP	Quantum critical point
QO	Quantum oscillations
RKKY	Ruderman–Kittel–Kasuya–Yoshida
SdH	Shubnikov–de Haas
SS	Surface state

STM	Scanning tunneling microscopy
TKI	Topological Kondo insulator
TRIM	Time-reversal invariant momentum
TRS	Time-reversal symmetry
WAL	Weak antilocalization
WF	Wiedemann–Franz

Chapter 1
Introduction

1.1 Kondo Effect

Contrary to free electron gases, strong electronic correlation in many-body systems often results in a wide variety of electronic ground states, leading to phenomena such as heavy Fermi liquids, unconventional superconductivity, magnetism, hidden order, strongly correlated insulators. Understanding those exotic electronic phases and excitations (i.e., emergent quasiparticles) from those ground states is the most fundamental and important research problem in condensed matter physics. Among those strongly correlated electronic systems, the physical properties of rare-earth compounds have been intensively studied. The key ingredient is the inner $4f$ electrons of rare-earth elements; these electrons are mostly localized in their atomic environment, and their moments play a role in magnetism. However, once the localized moments are immersed in a sea of mobile conduction electrons, they interact strongly on decreasing the temperature. Consequently, the localized moments can occasionally become itinerant through the interaction with the conduction electrons.

The problem of the interactions between localized moments and conduction electrons can be traced to the well-known Kondo problem, which concerns the resistivity minimum found in metals with dilute magnetic impurities. Hereafter, we refer to the problem as the dilute Kondo effect to distinguish it from the problem in a lattice system, which will be introduced in the next Sect. 1.2. The theoretical explanation for this phenomenon was first reported by Kondo in 1964 [1]. Kondo showed that the coupling J between conduction electrons and a magnetic moment yields singularity in the second perturbation scattering term, leading to a logarithmic increase in the resistivity through the factor -log T with decreasing temperature T. Corresponding to this singularity, the Kondo interaction involves the logarithmic growth of the effective interaction J:

$$J \rightarrow J(T) = J + 2J^2 N(0) \left(\ln \frac{D}{T} \right), \tag{1.1}$$

Y. Sato, *Quantum Oscillations and Charge-Neutral Fermions in Topological Kondo Insulator YbB*$_{12}$, Springer Theses,
https://doi.org/10.1007/978-981-16-5677-4_1

where $N(0)$ is the density of states (DOS) at the Fermi level and D is the width of the conduction band. It is worth noting that this term originates from both the noncommutative relation of the local moment spin operators, $[S_+, S_-] \neq 0$, and the Fermi–Dirac distribution function; therefore, it has roots in the quantum many-body effect. This formulation is quite novel in that the interaction becomes larger as the energy scale becomes smaller. The logic later sparked a novel concept called asymptotic freedom in particle physics. The logarithmic growth, however, implies that some physical properties cannot avoid divergence as $T \to 0$. This failure of the theory at low temperature is clear because the correction term becomes comparable to the non-perturbation term below the characteristic Kondo temperature T_K defined by

$$T_\mathrm{K} \sim D \exp\left[-\frac{1}{2JN(0)}\right]. \tag{1.2}$$

Therefore, the low-temperature physics below T_K remained a puzzle until the 1970s, stimulating further theoretical investigations. Wilson later developed the renormalization group theory and solved this problem [2]. Now, the Kondo physics is strictly solved throughout the temperature range [3]. The same expression of T_K can be also obtained from the scaling theory developed by Anderson [4]. In the strong coupling regime at low temperature, the spins of the localized moment and the sea of conduction electrons form a Kondo singlet, where the moment is screened by the sea of conduction electrons, and the magnetic degree of freedom of the moment disappears.

1.2 Kondo Lattice

As rare-earth compounds contain the "magnetic impurity" of $4f$ electrons at each lattice site, one can expand the above dilute Kondo problem by including lattice periodicity in the moments. The underlying physics of the system can be described by the Kondo lattice Hamiltonian:

$$H = -t \sum_{(i,j)\sigma} (c^\dagger_{i\sigma} c_{j\sigma} + \mathrm{h.c.}) + J \sum_{j,\alpha\beta} \boldsymbol{S}_j \cdot c^\dagger_{j\alpha} \boldsymbol{\sigma}_{\alpha\beta} c_{j\beta}, \tag{1.3}$$

where t is a hopping integral, $c^\dagger_{i\sigma}$ is an electron creation operator at site i with spin σ, and $\boldsymbol{S}_j$ is a spin operator of localized moments at site j. In the lattice system, the Kondo effect coherently occurs over the whole lattice, leading to the ground state of the non-magnetic Kondo singlet. This is because elastic scatterings at the local moment conserves momentum owing to translational symmetry [5]. Consequently, the system develops Fermi liquid (FL) behavior with the renormalized effective mass m^*, reflecting the strong electronic correlation. In Kondo lattice systems, m^* typically becomes 100–1000 times larger than that of a bare electron; hence, the quasiparticle is called a heavy fermion. Although all the physical properties show universal behavior with respect to T/T_K in the dilute Kondo system, this is not the case for a lattice system, which has another energy scale related to the Kondo coherence: $k_B T^*$. Here, T^* is called the coherent temperature below which the sea of conduction electrons requires coherence, leading to the formation of FL. T^* is always lower than T_K, below which the Kondo singlet locally begins to occur. The relation between T^* and T_K is a subtle problem, and it has been pointed out that T^* strongly depends on the density of conduction electrons [6].

There also exists another type of interaction between conduction and local electrons, known as the Ruderman–Kittel–Kasuya–Yoshida (RKKY) interaction. Immediately after a conduction electron interacts with a local moment, the spin of the conduction electron has polarization for a moment. As other local moments can experience this polarization through the conduction electron, one can regard this effect as an effective exchange interaction between two local moments. The RKKY interaction is given by

$$H_{\mathrm{RKKY}} = -9\pi \frac{J^2}{\epsilon_F} \left(\frac{N_e}{N} \right)^2 f(2k_F R) \boldsymbol{S}_1 \cdot \boldsymbol{S}_2, \tag{1.4}$$

where R is the distance between given two local moments and $f(x) = (-x \cos x + \sin x)/x^4$ is the function that describes the spatial modulation of the interaction. Because $f(x)$ oscillates by changing its sign as a function of R, the effective interaction becomes either ferromagnetic or antiferromagnetic. Now, the RKKY interaction is established as a major mechanism for the magnetism in heavy fermion systems.

The competition between the Kondo effect and RKKY interaction was pointed out by Doniach [7]. In the Doniach phase diagram shown in Fig. 1.1, the Kondo temperature T_K (the red curve) and Néel temperature T_N (the blue curve), which corresponds to the strength of the RKKY interaction, are plotted as a function of the coupling constant J. In the weak coupling regime of $T_N > T_K$, the ground state with magnetism becomes stable. In the strong coupling regime of $T_K > T_N$, on the other hand, the Kondo interaction overcomes the RKKY interaction, and Kondo screening leads to the coherent heavy fermion quasiparticles below T^*. In other words, the coupling strength J determines whether the f-electrons become localized or itinerant.

Around the point where $T_N \sim T_K$, the system undergoes a phase transition between these states; remarkably, the phase transition can occur even at $T = 0$. Most of the phase transitions occurring at finite temperature are driven by thermal fluctuation. On the other hand, the phase transition at absolute zero temperature is called a quantum phase transition, which is driven by the uncertainty principle of quantum mechanics. In the vicinity of the critical constant J_c, which is called the quantum critical point (QCP), novel quantum phenomena, such as unconventional superconductivity and non-FL behavior are occasionally realized. Several types of QCPs including heavy fermion systems have been reported thus far, and they have provided interesting platforms to study a series of exotic electronic states in strongly correlated systems [8].

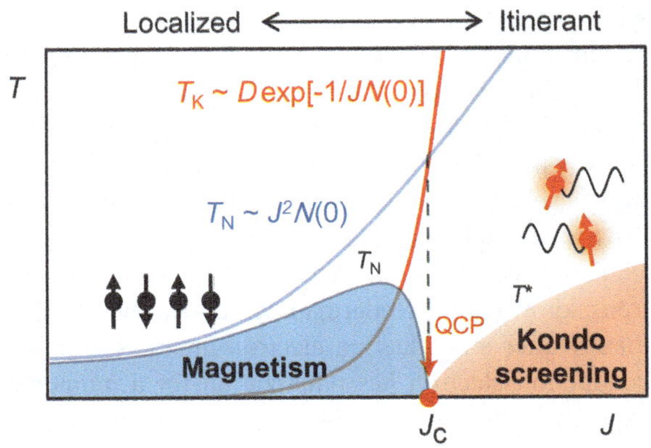

Fig. 1.1 Schematic of the Doniach phase diagram. Magnetic order and a Kondo screened state are realized according to the coupling strength J. f-electrons become itinerant on approaching the right side of the graph. © 2021 The author

1.3 Kondo Insulator

In the Kondo screening regime with a high coupling strength J, the renormalized coherent band dispersion of a heavy FL is well defined. However, the question of whether the system becomes metallic or insulating below the coherent temperature $T \ll T^*$ is not trivial. Several types of models have been proposed to study Kondo physics. To examine the insulating ground state, let us first express Eq. 1.3 in the limit of $t/J \to 0$ as

$$H = J \sum_{j,\alpha\beta} \mathbf{S}_j \cdot \boldsymbol{\sigma}_j + O(t), \tag{1.5}$$

where $\boldsymbol{\sigma}_j \equiv c^\dagger_{j\alpha} \boldsymbol{\sigma}_{\alpha\beta} c_{j\beta}$. The Hamiltonian simply describes the onsite antiferromagnetic ($J > 0$) interaction, leading to a non-magnetic ground state, where the Kondo singlet is formed at each site as

$$|\mathrm{KI}\rangle = \prod_j \frac{1}{\sqrt{2}} \left(\Uparrow_j \downarrow_j - \Downarrow_j \uparrow_j \right). \tag{1.6}$$

Here, the double and single arrows denote the spin of local and conduction electrons, respectively. It is apparent that this ground state has a spin gap of $2J$ as a triplet excitation and a charge gap of $3J$, which separates the hole and electron quasiparticle dispersion bands [5]. Therefore, the Kondo singlet state in the limit of $t/J \to 0$ is insulating. The Kondo screened system, in which the Fermi energy lies in the charge gap, is called a Kondo insulator (KI). In addition to Mott insulators, KIs are known to be a typical example of correlated insulators, which acquire the insulating ground state owing to strong electron correlation.

To examine how the Kondo lattice becomes either a KI or a heavy FL, it is instructive to introduce the band structure of the system. The nature of the insulating ground state can be understood by introducing the band hybridization, and this view

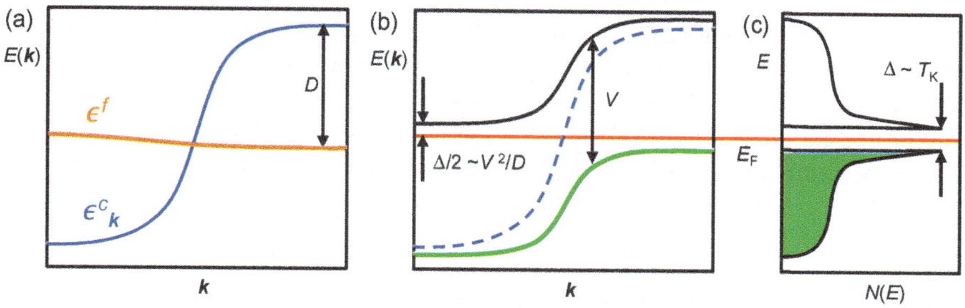

Fig. 1.2 **a** Schematic of a conductive and flat band without hybridization, realized in typical Kondo lattice compounds at a temperature above T_K. As the Fermi energy E_F crosses the conduction band, the system becomes metallic. **b** The hybridized band structure at a temperature below T_K. **c** Energy dependence of the DOS for the hybridized band structure with half filling. © 2021 The author

is preferable especially when we discuss the topological order of a certain insulator in Sect. 1.5. The model we employ here is the periodic Anderson model:

$$H = \sum_{k\alpha} \epsilon^c_{k\alpha} n^c_{k\alpha} + \epsilon^f \sum_{i\alpha} n^f_{i\alpha} + \sum_{k\alpha} \left(V_k f^\dagger_\alpha c_{k\alpha} + \text{h.c.} \right) + U \sum_{\alpha \neq \alpha'} n^f_{i\alpha} n^f_{i\alpha'} \tag{1.7}$$

Here, $\epsilon^p_{k\alpha}$ and $n^p_{k\alpha}$ are the energy band and number operator, where the corresponding superscript $p = c$ or f denotes c- or f-electrons, respectively. V_k and U denote the strength of the hybridization and onsite Coulomb repulsion interaction, respectively. The band structure for $V = |V_k| = 0$ is depicted in Fig. 1.2a. Based on its localized character of the f-moments, ϵ^f can be regarded as nearly flat. Furthermore, the system is metallic at a sufficiently high-temperature relative to T_K because the Fermi energy E_F lies within the conduction band. With decreasing temperature, the Kondo interaction becomes increasingly pronounced, as expressed in Eq. 1.1. When V becomes dominant, these two bands begin to hybridize, leading to the reconstructed band structure shown in Fig. 1.2b. Here, the dispersion has the direct gap V and indirect gap $\Delta \sim 2V^2/D$, where D is the half band width [9]. It has been shown that Δ is robustly finite against U [10]. On approaching to the limit of $U \to \infty$ adiabatically, the model is reduced to Eq. 1.5, leading to an insulating ground state with the finite charge gap $\Delta \sim k_B T_K$. When the system is not at half filling or the Fermi level is shifted upwards (downwards) by introducing a moderate amount of electrons (holes), a finite DOS can be induced at the Fermi level. In this case, the quasiparticle band has a much lower curvature than the original conduction band as the conduction electrons acquire an f-like character owing to the hybridization. A lower curvature of a band dispersion leads to the heavier effective mass of quasiparticles. On the other hand, when the system is at half filling, the hybridized band with a finite Δ leads to an insulating ground state; hence, the Kondo lattice becomes a KI.

Owing to the band inversion, the flat f-band splits into several patches, some of which are partially occupied in the hybridized band. This partial occupation of the f-orbit results in the intermediate valence (or mixed valence) state widely found in Kondo screened materials including KIs. Taking an example of Yb compounds, in the intermediate valence state, the atomic configuration of magnetic (Yb^{3+}) and non-magnetic (Yb^{2+}) states is coherently superpositioned spatially and temporally. Therefore, the mean valence for the Yb ion becomes a non-integer ranging from 2 to 3. The intermediate valence state is also one of the research interest in KIs.

1.4 Electronic Properties of Kondo Lattice Systems

1.4.1 Thermodynamic and Transport Properties

Figure 1.3a–c shows the temperature evolution of the magnetic susceptibility χ, resistivity ρ, and electronic specific heat C_e of a KI and heavy FL by red and blue curves, respectively. In the high-temperature regime $T \gg T_K$, these quantities show similar temperature dependences because both systems can be treated as conduction electrons strongly scattering with lattice local moments. χ shows Curie–Weiss behavior owing to unscreened local moments. The resistivity shows metallic behavior at a sufficiently high temperature, and it exhibits a logarithmic upturn below T_K. On decreasing the temperature further, the Kondo lattice acquires coherence below T^*, and the system becomes either a heavy FL or a KI according to the band structure and chemical potential.

First, we examine the electronic properties of a heavy FL. As the entropy S of the local moments $R \ln 2$ (here, R is the universal gas constant) vanishes below T^* owing to the Kondo screening, the C_e/T of the conduction electron is also influenced according to

$$S(T) = \int_0^T \frac{C_e(T')}{T'} dT'. \tag{1.8}$$

Therefore, on decreasing the temperature, C_e/T increases, as if the conduction electrons absorb the magnetic entropy of the local moments. To understand the physical picture behind this enhanced C_e, it is instructive to express the Sommerfeld coefficient $\gamma = C/T(T \to 0)$ of the 3D electron gas model as

$$\begin{aligned}\gamma &= \frac{\pi^2 k_B^2}{3} N(0) \\ &= \frac{k_B^2 k_F}{3\hbar^2} m^*,\end{aligned} \tag{1.9}$$

where k_B is the Boltzmann constant and k_F is the Fermi wave vector. As $\gamma \propto N(0) \propto m^*$, the enhancement in γ implies that the effective mass of the quasiparticle is larger than that of a bare electron. This is why the quasiparticles are called heavy fermions. According to Eq. 1.8, γ should be of the order of $\sim (R \ln 2)/T_K$ at sufficiently low temperatures. When the Fermi energy crosses the heavy fermion band, the system

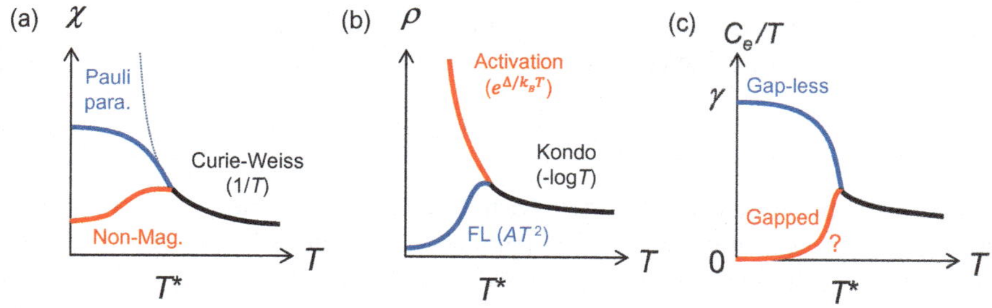

Fig. 1.3 Schematics of electronic properties expected in Kondo screened materials, namely the Kondo insulator (the red curves) and heavy FL (the blue curves). **a** Magnetic susceptibility χ, **b** resistivity ρ, and **c** electronic specific heat divided by temperature C_e/T are shown as a function of temperature. ©2021 The author

becomes metallic, and its physical quantities can be well described in the framework of FL theory. As we will see in Sects. 1.4.2 and 1.4.3, χ shows a temperature-independent Pauli paramagnetic response, while ρ decreases in proportion to A_2T^2 (the blue curves in Fig. 1.3a–c). These are the basic properties of a FL.

On the other hand, when the Fermi level is within the Kondo gap, Kondo lattice becomes a KI. It shows a weak magnetic response and activation behavior ($\propto \exp[\Delta/k_\mathrm{B}T]$) in resistivity, as indicated by the red curves in Figs. 1.3a–b. One of the main problems addressed in this thesis is the behavior of C_e/T in KIs at low temperature. From a conventional view of insulators, $N(0)$ becomes zero for $T \to 0$. In this case, fermionic excitations must derive from the thermally activated quasiparticles across the gap Δ, leading to $C_\mathrm{e}/T \sim \exp(-\Delta/k_\mathrm{B}T)$. However, if C_e/T behaves as same as in a heavy FL at high temperature, as indicated by the black curve in Fig. 1.3c, this conventional gapped behavior violates Eq. 1.8. Therefore, there may be an additional steep peak in C_e/T below T^*. Experimentally, however, reliably estimating C_e/T from the total specific heat is difficult, because other contributions, such as the contribution from phonons, become dominant at $T > T_\mathrm{K}$ (here, $T_\mathrm{K} \approx 100\,\mathrm{K}$ in typical KIs). In fact, the thermodynamics of KIs is not fully understood, and as we will discuss in Chap. 5, every model KI discovered so far, to our knowledge, exhibits finite γ. This is quite surprising given that they show thermally activated behavior in ρ, i.e., they are charge insulators with zero DOS at the Fermi level.

1.4.2 Kadowaki–Woods Ratio

The applicability of the FL theory to heavy fermions implies that the quasiparticles in Kondo lattice systems behave as non-interacting free particles but with a heavier effective mass owing to the renormalization of the Kondo interaction. This picture leads to the universal properties of FLs for various families of materials. The temperature dependence of resistivity, for example, is given by

$$\rho = \rho_0 + A_2 T^2, \tag{1.10}$$

where ρ_0 is the impurity scattering term and the T^2 dependence derives from electron–electron interaction with the parameter A_2, which determines the probability of the scattering. Kadowaki and Woods [11] first pointed out the universal relationship between γ and A_2:

$$\frac{A_2}{\gamma^2} = R_{\rm KW} \equiv 1 \times 10^{-5} \mu\Omega{\rm cm}({\rm Kmol/mJ})^2. \tag{1.11}$$

Here, $R_{\rm KW}$ is known as the Kadowaki–Wood (KW) ratio. Although each physical quantity varies by several orders of magnitude among various classes of materials, this relation holds in a rich variety of systems, such as normal metals, f- and d-electron-based intermetallic compounds, and oxides. The relation can be understood if one considers the simplest expressions for A_2: $A_2 \propto m^{*2}$. Because $\gamma \propto m^*$ from Eq. 1.9, the ratio $R_{\rm KW}$ does not depend on the renormalization parameter m^*, leading to the universal ratio. Although it is also known that the ratio A_2/γ^2 in a certain category of materials becomes 0.04 times smaller than $R_{\rm KW}$, the deviation can be generalized by considering the degeneracy of quasiparticles [12].

1.4.3 Wilson Ratio

One can also find a universal relationship between γ and the magnetic susceptibility χ for FLs. In the FL theory, the main contribution to magnetism is from the Zeeman splitting of the spin-up and spin-down bands, which leads to unbalanced spin population phenomena known as Pauli paramagnetism. This susceptibility is given by

$$\chi_{\rm Pauli} = \frac{(g\mu_{\rm B})^2}{4} N(0), \tag{1.12}$$

which is temperature independent, in contrast to the Curie–Weiss behavior. Here, g is the g-factor, and $\mu_{\rm B}$ is the Bohr magneton. Pauli paramagnetism is known as a basic property that metals exhibit at sufficiently low temperatures below the Fermi temperature. According to Eq. 1.9, $\gamma \propto N(0)$ leads to the universal ratio between $\chi_{\rm Pauli}$ and γ:

$$R_{\rm W} = \frac{3}{4}\left(\frac{\mu_{\rm B} g}{2\pi k_{\rm B}}\right)^2 \frac{\chi_{\rm Pauli}}{\gamma}, \tag{1.13}$$

which is known as the Wilson ratio [2]. It has been shown that $R_{\rm W}$ is close to unity for many FL materials. It is quite surprising that, despite the rich diversity of ground states in correlated metals, the low-temperature physical properties are solely determined by the degree of the renormalization parameter $N(0) \propto m^*$. Therefore, the universal relationships we have discovered thus so far, the KW ratio and Wilson ratio, have been considered direct and fundamental evidence for the validity of the FL theory for various correlated metals.

1.4.4 Wiedemann–Franz Law

In a FL, quasiparticles carry not only charge but also heat. Therefore, one can expect a quantitative relation between charge transport and thermal transport. To find such a relation, we first express the electric conductivity σ_{xx} in a 3D electron gas model is based on the Drude model as

$$\sigma_{xx} = \frac{ne^2\ell}{m^* v_{\mathrm{F}}} \tag{1.14}$$

$$= \frac{e^2}{3\pi^2\hbar}\ell k_{\mathrm{F}}^2, \tag{1.15}$$

where $n = k_{\mathrm{F}}^3/3\pi^2$ is the carrier density, e is the elementary charge, ℓ is the mean free path, and $v_{\mathrm{F}} = \hbar k_{\mathrm{F}}/m^*$ is the Fermi velocity. The thermal conductivity κ of the quasiparticle is also expressed as

$$\kappa_{xx} = \frac{1}{3}\gamma v_{\mathrm{F}}\ell T. \tag{1.16}$$

By substituting Eq. 1.9, one can obtain

$$\frac{\kappa_{xx}}{T} = \frac{k_{\mathrm{B}}^2}{9\hbar}\ell k_{\mathrm{F}}^2. \tag{1.17}$$

Combining Eqs. 1.15 and 1.17 yields

$$\frac{\kappa_{xx}/T}{\sigma_{xx}} = \frac{\pi^2}{3}\left(\frac{k_{\mathrm{B}}}{e}\right)^2 \equiv L_0 \simeq 2.44 \times 10^{-8}\,\mathrm{W\Omega/K^2}. \tag{1.18}$$

The universal constant L_0 is known as Lorentz number, and the relation given by Eq. 1.18 is known as Wiedemann–Franz (WF) law [13]. This is valid for not only the longitudinal transport component but also the transverse response, i.e., $\kappa_{xy}/\sigma_{xy}T = L_0$. As this law holds for most conventional metals, it is regarded as supporting evidence for the validity of the FL theory. However, it is worth noting that the charge-neutral excitations can contribute to the thermal transport, while they cannot contribute to the electrical conductivity. Indeed, the phonon, which is an emergent excitation mode of lattice vibration, is a typical neutral quasiparticle that can carry heat. Therefore, if the observed ratio κ/σ is larger than the Lorentz number, it can be interpreted that other neutral quasiparticles contribute to the heat transport. In fact, there are several examples that violate the WF law in strongly correlated electron systems such as quantum spin liquids [14, 15] and high-T_c cuprates [16]. In addition, the violation of the WF law has also been reported in the heavy fermion $YbRh_2Si_2$, solely near the magnetic QCP [17, 18], where the κ/σ becomes smaller than L_0. As neutral excitation descriptions are not valid in this case, the anomalous electric transport property is considered to be linked to the non-FL behavior, the mechanism behind which is another central issue in correlated materials [19, 20]

1.5 Topological Kondo Insulator

The application of the concept of topology to the field of condensed matter physics led to a paradigm shift in research of a new class of materials [21, 22]. A key idea arises from the question of whether it is always possible to adiabatically deform the given wave function of an insulator to another. The answer is no: in the presence of both time-reversal symmetry (TRS) and inversion symmetry (IS), all band insulators can be classified into two with the Z_2 invariant $\nu = 0, 1$, and these electronic structures are topologically distinct. The criteria for this classification are given by the Fu–Kane formula [23]:

$$(-1)^{\nu} = \prod_i \zeta(\Lambda_i) = \begin{cases} +1 \text{ trivial (conventional insulator)}, \\ -1 \text{ non-trivial (topological insulator)}, \end{cases} \tag{1.19}$$

where $\zeta(\Lambda_i)$ is the parity eigenvalue of the occupied Bloch states at the time-reversal invariant momentum (TRIM) Λ_i. As the wave functions of TIs show twisting, these two classes of insulators cannot be connected without breaking some symmetries or closing the band gap. As one can regard vacuum as a conventional insulator, this classification forces the band gap to close at the boundary between TIs and vacuum (i.e., sample edge or surface). This is called bulk-edge correspondence. Hence, the existence of topological order is manifested by the metallic surface state (SS) surrounding the insulating bulk. Because this SS is protected by the symmetry of the underlying crystal, it is robust to perturbations that do not violate the crystalline symmetries. The remarkable feature of this SS is that its dispersion is described as a linear Dirac cone of mass-less fermions. In addition, TRS and IS guarantee band crossing with opposite spin directions at TRIM. This constraint leads to spin-momentum locking, where the spin direction is determined by the momentum of the electron. Furthermore, the backscattering in this surface channel is prohibited owing to the topological protection. Because of the robustness, spin-momentum locking, and topological nature of SS, TIs recently attracted much interest for possible future applications such as efficient spintronics devices, quantum computation, and memory storage with low energy consumption [24].

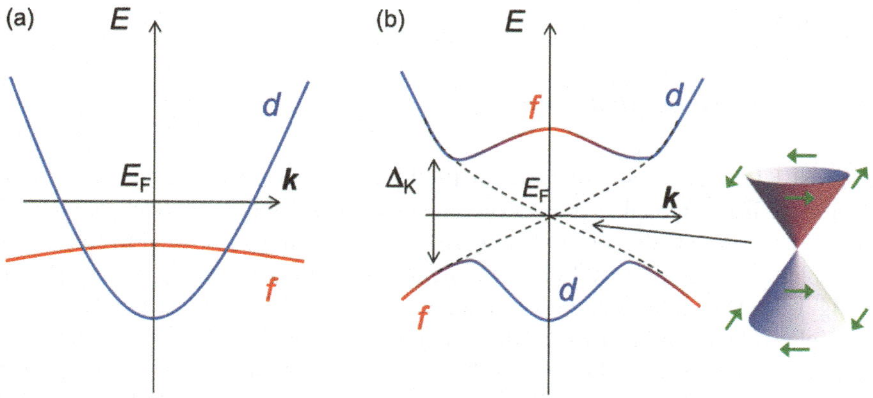

Fig. 1.4 Schematics of the band structure of KIs at a **a** high T and **b** low T. The dotted lines indicate surface states with Dirac dispersion, where spins of electrons are locked to their momentum. © 2021 The author.

The integer quantum Hall insulator is a prototypical example of this topological phase of matter, and Z_2 TIs were later discovered mostly in materials with strong spin-orbit interaction. In the early stage, the study of TIs focused on materials with weak correlation, most likely because their theoretical treatment to calculate band structures is relatively easy. In 2010, Dzero et al. pointed out that a certain type of KI can be topologically non-trivial because they inherently experience band inversions owing to the hybridization of bands with opposite parity [25]. Owing to this new scheme, KIs have been intensively re-investigated as a new family of TIs, where the strong electron correlation drives topological order. These systems are called topological KIs (TKIs) [10].

Because the topological order in KIs is driven by Kondo effect, its topological order also evolves with temperature. Figure 1.4a schematically shows a simplified band structure of KIs at high temperature ($T > T^*$). As the conduction d-band crosses the Fermi level E_F, the system becomes metallic with a flat f-band. As the temperature decreases to $T < T^*$, the d-orbital begins to create a coherent heavy quasiparticle band through the hybridization with the flat f-band, as shown in Fig. 1.4b. Because the conduction d-band and localized f-band are even and odd in the parity operation, respectively, this hybridization leads to band inversions with different parities, possibly at several TRIMs. According to Eq. 1.19, therefore, if the band inversions occur an odd number of times over the whole Brillouin zone, KIs can be topologically non-trivial. Consequently, the KI can host a metallic SS on its surface, which is protected by TRS and IS, as shown in Fig. 1.4b. In Sects. 1.6 and 1.7, we review the electronic properties of candidate TKI materials, SmB_6 and YbB_{12} and discuss experimental evidence for the existence of the topological SS.

1.6 SmB_6

1.6.1 Electronic Structure

The first KI, SmB_6, was discovered in 1969 [26], and its electrical properties have been intensively studied as a model KI materials and, later, as a correlation-driven TI [27]. It forms a body-centered cubic structure consisting of Sm and B_6 octahedra with a lattice constant $a \approx 4.133$ Å, belonging to the $Pm3m$ space group, as shown in Fig. 1.5a. The corresponding Brillouin zone (BZ) is also shown in Fig. 1.5b. First principles calculations using local density approximation (LDA) with the Gutzwiller method revealed that band inversions between the d-band and f-band are expected to occur at three $\bar{X}$ points [28, 29]. Therefore, according to the Eq. 1.19, these band inversions force SmB_6 to be a Z_2 non-trivial insulator. The resulting band dispersion of the SS is shown by the red curves in Fig. 1.5c, and the inset depicts corresponding Fermi surfaces on the (001) surface. Given that most TIs with strong spin-orbit coupling have a single Dirac cone near the Fermi level [21], it is notable that multiple Fermi surfaces appear in SmB_6 at the $\bar{\Gamma}$ point and two $\bar{X}$ points. Furthermore, the SS is within the gap and is well separated with the bulk bands, which is favorable to study the intrinsic transport properties of a topological SS.

The formation of the hybridization gap in SmB_6 has been confirmed using various methods. For example, the insulating thermal activation behavior in resistivity has been confirmed [26, 30, 31], and it yields a gap amplitude $\approx$4 meV [31]. Figure 1.6a shows the Arrhenius plot of $\ln \rho$ versus $1/T$ [30]. At high temperatures in the range of 100–300 K, SmB_6 shows bad metallic behavior because the conduction electrons strongly scatter with the local moments owing to the absence of the coherent hybridization. As the temperature decreases, the resistivity gradually shows insulating behavior down to $\approx$4 K. It is well known that the resistivity shows saturation at low temperature, which is now discussed in terms of topological SSs. The details of the SS detected by transport measurements will be discussed in Sect. 1.6.2. The formation

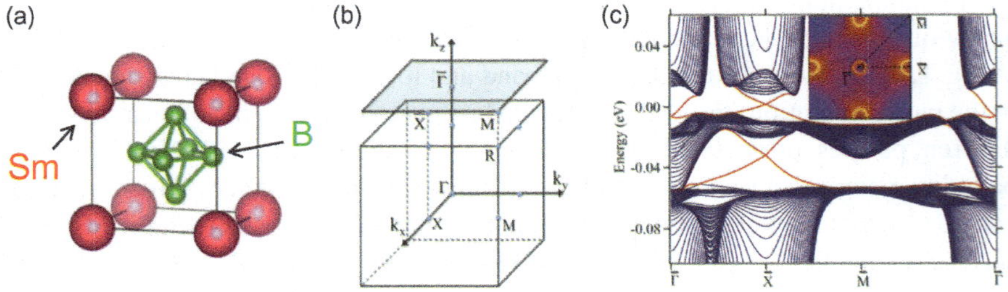

Fig. 1.5 **a** Crystal structure of KI SmB_6. © 2021 The author. **b** The bulk and (001) surface Brillouin zone of SmB_6. **c** Calculation of the surface and bulk band structures indicated by the red and dark purple curves, respectively. The inset shows Fermi surfaces of the topological SS on the (001) surface. **b** and **c** are reprinted figures with permission from [28] © 2013 by the American Physical Society

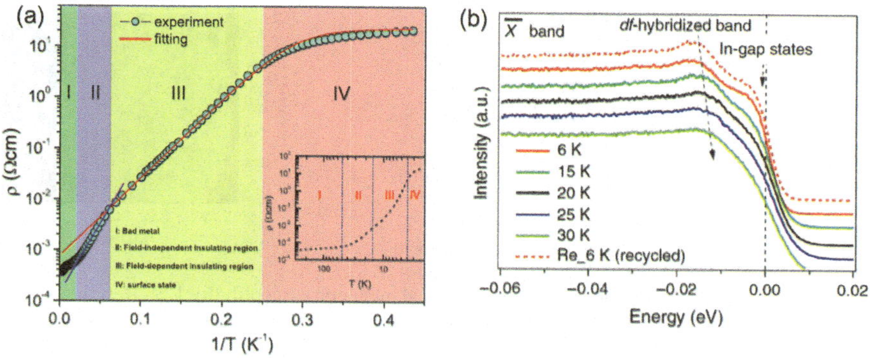

Fig. 1.6 Formation of the hybridization gap in SmB_6. **a** Arrhenius plot of ln ρ versus $1/T$. Reprinted figure with permission from [30] © 2015 by the American Physical Society. **b** Momentum-integrated ARPES spectral intensity at the $\bar{X}$ band. Reprinted by permission from Springer Nature: Nature Communications [35], © (Copyright Springer Nature Singapore Pte Ltd. 2013)

of the hybridization gap can be also observed from the suppression in DOS across the Fermi level by spectroscopy measurements such as scanning tunneling spectroscopy (STM) [32, 33], point-contact spectroscopy [34], and angle-resolved photoemission spectroscopy (ARPES) [35–37]. It is reported that the tunneling conductance $\mathrm{d}I/\mathrm{d}V$, which reflects the local DOS, shows gap-like feature around $V = 0$ and it develops below 100 K [32]. Momentum-integrated ARPES data are displayed in Fig. 1.6c. The peak structure resulting from the hybridization, which evolves with decreasing temperature, is clearly resolved. Recent ARPES measurements also revealed an in-gap state within the gap, the origin of which is still controversial [38].

SmB_6 also shows behavior characteristic of KI in its magnetic properties. Figure 1.7a shows the reciprocal magnetic susceptibility $1/\chi$ as a function of temperature [33]. At high temperature, the susceptibility shows paramagnetic Curie–Weiss behavior with a paramagnetic Curie temperature ≈50 K, which is derived from the local moments of f-electrons. $1/\chi$ gradually deviates from the T-linear behavior below ≈120 K, where the hybridization gap is formed. This Kondo screening prevents magnetic ordering, and χ shows less T-dependent non-magnetic behavior. χ also exhibits a low-temperature anomaly, which is likely to be associated with the in-gap state. Nuclear magnetic resonance (NMR) measurement also reveals the macroscopic magnetic properties of materials. The Knight shift K is related to the local static susceptibility χ_0 by $K \approx A/\gamma_e\gamma_n\hbar \cdot \chi_0$, where γ_e and γ_n are the electron and nuclear gyromagnetic ratio, respectively. The reported ^{11}B NMR Knight shift K also shows a similar temperature dependence as the bulk susceptibility χ [39]. NMR measurements are also used to obtain information on DOS $N(0)$ from spin-lattice relaxation rates $1/T_1$ by

$$\frac{1}{T_1} \propto N(0)^2 k_\mathrm{B} T. \tag{1.20}$$

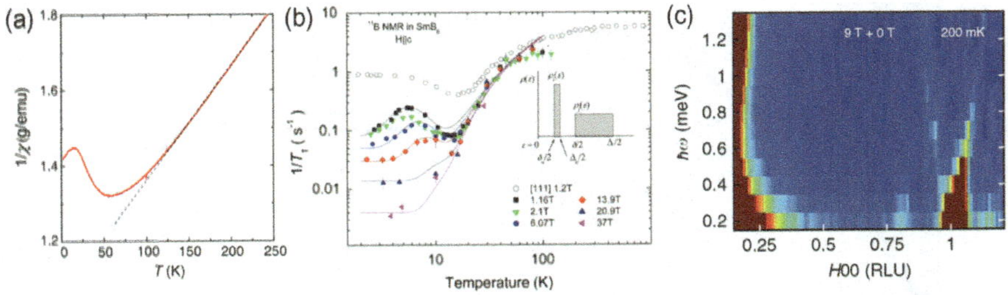

Fig. 1.7 **a** Temperature dependence of the reciprocal magnetic susceptibility $1/\chi$. Reprinted figure with permission from [33] © 2014 by the American Physical Society. **b** Temperature dependence of spin-relaxation rates $1/T_1$ in various fields. Reprinted figure with permission from [39] © 2007 by the American Physical Society. **c** Neutron scattering intensity mapped with $\hbar\omega$ versus the reciprocal vector H. Reprinted from [40] by The Author licensed under CC BY 4.0

$1/T_1$ in Fig. 1.7b shows suppression below $\approx$100 K, which is consistent with the gap opening. $1/T_1$ becomes T-independent at low temperatures, whereas it remains sensitive to the magnetic field. This anomaly at low temperature also implies a field-dependent in-gap state. Neutron scattering measurements were also conducted to detect magnetic excitations in SmB_6. Only residual Bragg scattering and weak phonon scattering were observed, and no magnetic ordering or excitations were resolved down to 200 mK, as shown in Fig. 1.7c [40]. These experiments reveal temperature-driven Kondo screening below $T \approx$100 K and the non-magnetic ground state in SmB_6.

1.6.2 Topological Surface State

The electronic structure of SmB_6 is well described in a Kondo screening model and has been regarded as a model KI material, with the exception of the low-temperature anomaly of the in-gap state. We will discuss the topological SS as a possible source of this contribution.

Numerous transport studies on SmB_6 were reported in an effort to detect the topological SS from different perspectives, such as the doping effect [41], geometrical effect [31, 42, 43], weak antilocalization (WAL) effect on magnetoresistance (MR) [44], spin polarization [45], and microcracks [46]. First, it was proved that the resistance plateau is sensitive to magnetic impurities [41]. As the SS is protected by TRS, it is expected to be destroyed by a magnetic impurity that violates TRS. As shown in Fig. 1.8a, the resistance plateau shows different behaviors depending on whether the impurity is magnetic. The introduction of 3% Gd is sufficient to destroy the SS, and the resistivity shows diverging behavior toward $T \to 0$. In contrast, Y-doping does not affect the transport properties.

It has also been shown that the low-temperature transport properties are sensitive to the geometry of samples and contacts. The upper panel of Fig. 1.8b displays

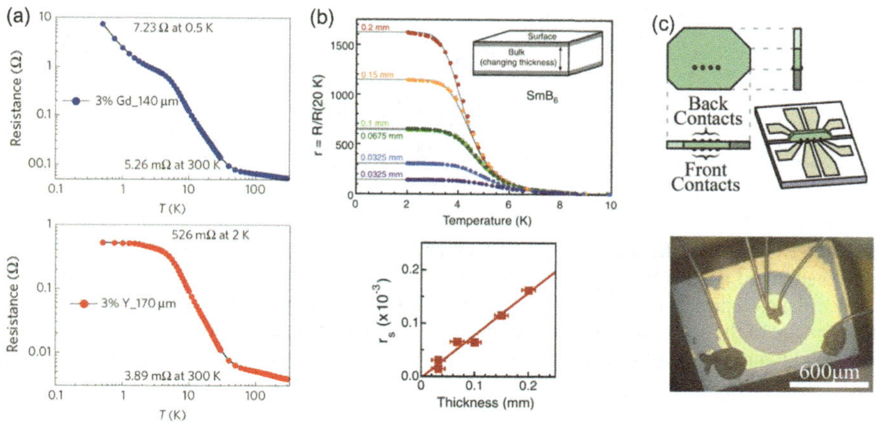

Fig. 1.8 Topological SS detected using several types of transport techniques. **a** Temperature dependence of resistivity of Gd-doped (upper panel) and Y-doped (lower panel) SmB_6 crystals. Reprinted by permission from Springer Nature: Nature Materials [41], © (Copyright Springer Nature Singapore Pte Ltd. 2014). **b** Temperature dependence of normalized resistivity with various thickness (upper panel). The lower panel displays the residual resistivity against thicknesses. Reprinted figure with permission from [42] © 2015 by the American Physical Society. **c** Examples of special configurations of contacts on the crystal. Reprinted figures with permission from [43, 46] © 2013 and 2020 by the American Physical Society

the resistivity data with various values of the sample thickness t [42]. The residual resistivity $\rho(T \to 0)$ shows a decreasing tendency with decreasing sample thickness. The lower panel shows the relationship between the residual resistance and sample thickness. To explain this behavior, one can regard the TI as a parallel circuit of a 3D bulk and 2D surface (inset of 1.8b) as follows:

$$\frac{1}{\rho} = \frac{\ell}{wt}\left(\frac{1}{R_{3D}} + \frac{1}{R_{2D}}\right) = \frac{1}{\rho_{3D}} + \frac{1}{\rho_{2D}} \cdot \frac{1}{t}. \tag{1.21}$$

As $1/\rho_{3D} \to 0$ for $T \to 0$, the saturation amplitude corresponds to the second term. Therefore, the linear relationship between the residual resistivity and t provides evidence for a 2D conduction channel in this compound. The separation of the surface and bulk conductance can also be achieved by employing specific configurations of the contacts [43], such as a double-sided Corbino disk [46]. Figure 1.8c shows some such configurations, where the contacts are made from both sides of the crystal. Transport measurements agree with simulation for this configuration, revealing the surface conduction in SmB_6. These techniques can also be employed as a powerful tool for detecting bulk conductance even in the surface-dominant regime.

Now, the question is whether the 2D conduction originates from the topological nature. One defining character of the topological SS is that the surface electrons have spin texture, where the spin is locked to its momentum. The detection of this spin polarity using a transport method is important as a possible future device application.

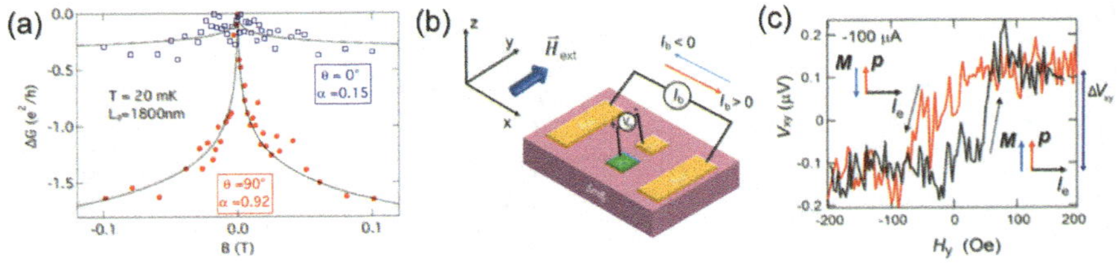

Fig. 1.9 Evidence for the spin-momentum locking in the SS of SmB_6. **a** Magnetoresistance under a magnetic field applied perpendicular (the red points) and parallel (the blue points) to the major surface plane. Reprinted figure with permission from [44] © 2016 by the American Physical Society. **b** Schematic of the setup for the spin potentiometric measurement and **c** the resulting spin voltage. Reprinted figure with permission from [45] © 2019 by the American Physical Society

The coupling between the spin and its momentum can be manifested in the WAL effect, which is a negative quantum correction to the conductivity. Applying magnetic fields destroys this coherence and leads to a cusp-like feature on MR. Figure 1.9a shows the low-field MR of SmB_6 with a magnetic field parallel and perpendicular to the conduction surface [44]. The cusp-like feature is more pronounced when the field is applied perpendicular to the surface plane, which is consistent with the framework of the WAL effect. This result, therefore, implies that the spin of SS is locked to its momentum. A more detailed discussion in this point will be given in the Chap. 4.

One can probe the spin polarization of the conduction electron more intuitively through spin potentiometric measurements [45]. In such measurements, the contact to the sample is made in a conventional Hall bar geometry, but one of the contacts is replaced by a ferromagnetic metal. As illustrated in Fig. 1.9b, a current is induced between the gold electrodes on both sides of the crystal, and the transverse voltage is measured between the gold and permalloy magnetic contact (the green region Py). By changing the direction of magnetization of the Py contact, one can also change the potential in different spin directions. The spin-voltage graph shows a finite hysteresis loop in the transverse voltage V_{xy} and the field H_y along the y-direction. This result provides evidence that the spin polarization is perpendicular to the current (momentum), which is again consistent with the spin texture we can expect in the topological SS.

More direct evidence for the spin texture in the topological SS band in SmB_6 was provided by spin ARPES measurement [47]. A schematic of the Fermi surfaces on the (001) surface and their spin texture is illustrated in Fig. 1.10a. The spin of the β bands located on $\bar{X}$ rotates counterclockwise. The β bands are clearly resolved by the ARPES measurement, as shown in Fig. 1.10b, and the measurement shows good agreement with theoretical calculations. The α band is also resolved at the center of the $\bar{\Gamma}$ point, but the intensity is smaller than that of the β bands. Figure 1.10c displays the ARPES intensities along the C1 cut indicated by the red line in Fig. 1.10b for each spin direction. The up and down spins along the x direction show a significant

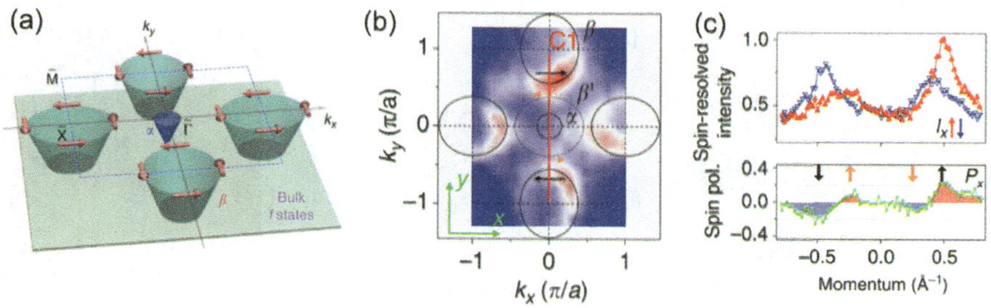

Fig. 1.10 Spin texture of the topological SS in SmB_6. **a** Schematics of the SS on the (001) surface. **b** Observed Fermi surfaces of the SS. The red line C1 in the middle is the cut along which the spin polarity is measured. **c** Spin-resolved intensity and spin polarization along the C1 cut. Reprinted by permission from Springer Nature: Nature Communications [47], © (Copyright Springer Nature Singapore Pte Ltd. 2014)

difference, and the spin polarization is consistent with the theoretically predicted spin texture. This result provides direct evidence that the SS originates from the non-trivial topology in the bulk electronic state.

1.7 YbB_{12}

1.7.1 Electronic Structure

YbB_{12} is another prototypical KI. The study on poly-crystalline powders started in the 1970s [48, 49], and the first composition of large single crystals was achieved using the floating zone method in 1998 [50]. The crystalline structure of YbB_{12} is shown in Fig. 1.11a. It has a face-centered cubic structure consisting of Yb and B_{12} cubooctahedra with the lattice constant $a \approx 5.28$ Å, belonging to the $Fm\bar{3}m$ space group. The Brillouin zones of the bulk and the (001) surface are shown in Fig. 1.11b. The topological invariant for this material has also been calculated, but it was shown that the Z_2 invariant in YbB_{12} is trivial ($\nu = 0$) because the band inversion occurs twice at three X points [51]. However, one can still expect a topological SS protected by a crystalline symmetry such as mirror symmetry. A topological non-trivial phase protected by a mirror symmetry is called a topological crystalline insulator (TCI) [52], which was first proposed in the strong spin-orbit coupling system $Bi_{1-x}Sb_x$. As the mirror operation classifies the wave function into two subgroups according to its eigenvalue $\eta = \pm i$, one can calculate the Chern number $N_{\pm i}$ for each subgroup. This is analogous to the fact that the TRS yields two subgroups according to the spin, $\pm 1/2$. TCI is a topological phase having a nonzero mirror Chern number, which is defined as $N_M = N_{+i} - N_{-i}$, but zero total Chern number $N = N_{+i} + N_{-i}$. In YbB_{12}, it was shown that $N_M = 2$ (non-trivial) owing to the presence of the (010) mirror plane [51]. Therefore, it is expected that YbB_{12} also hosts a topological SS. The result of

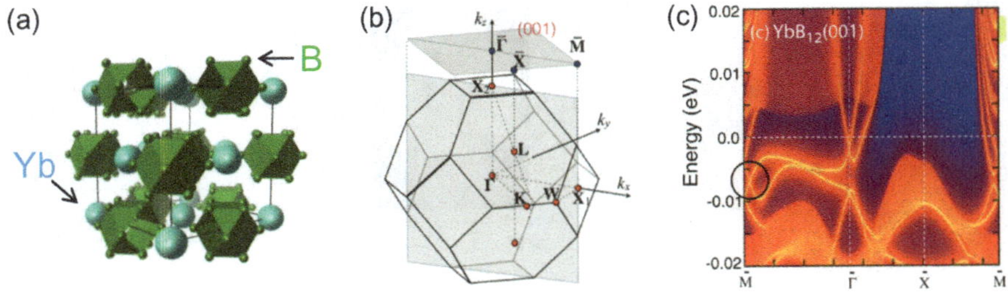

Fig. 1.11 **a** Crystal structure of KI YbB_{12}. © 2021 The author. **b** Bulk and (001) surface Brillouin zones of YbB_{12} and the calculation of the (001) surface band. The black circle in **c** marks the Dirac point protected by mirror symmetry. **b** and **c** are reprinted figures with permission from [51] © 2014 by the American Physical Society.

first principles calculations for the (001) surface band is displayed in Fig. 1.11c. As the mirror plane passes through $\bar{M}$-$\bar{\Gamma}$ direction, the degeneracy of the two substates are guaranteed on this plane. Therefore, the Dirac point indicated by the black circle in Fig. 1.11c is topologically protected by the mirror symmetry. The surface Dirac cone protected by the finite mirror Chern number also hosts a spin texture as well as an ordinal TI [53].

As a KI, the physical properties of YbB_{12} are quite similar to those of SmB_6. The Arrhenius plot of its resistivity is displayed in Fig. 1.12a. YbB_{12} develops multiple energy gaps: $\Delta E_1 \sim 15$ meV below $T \approx 40$ K and $\Delta E_2 \sim 4$ meV below $T \approx 15$ K [50]. The low temperature plateau is also observed in YbB_{12}, and it can be interpreted as a conduction channel of the topological SS. The transport properties in YbB_{12} will be described with a great attention in Chap. 4. The formation of the second gap is unique in YbB_{12}, and its origin is still under the debate. The magnetic susceptibility χ also shows the typical Kondo screening behavior [54]. The Curie–Weiss behavior is observed at high temperatures, and χ shows a peak, which is suppressed below $T^* \approx 80$ K owing to the Kondo screening and finally becomes less T-dependent on decreasing the temperature further. The substitution of Yb by Lu moves the system towards a normal metal because Lu does not contain $4f$ electrons. The substitution strongly suppresses the Curie–Weiss behavior, and the system gradually shows Pauli paramagnetic behavior towards the Lu end. This result clearly demonstrates that the $4f$ electrons in Yb ions play an important role in its electronic properties and the observed behavior is entirely consistent with the Kondo physics. This behavior is also supported by the NMR Knight shift [55, 56], as shown in Fig. 1.12(c).

The formation of the hybridization gap was also observed through optical conductivity [57] and photoemission spectroscopy (PES) measurements [58, 59]. At high temperatures, optical reflectivity $R(\omega)$ shows rather metallic behavior of plasma reflection as $R(\omega = 0) \rightarrow 1$. However, at low temperatures, $R(\omega)$ shows a peak feature at $\omega \sim 15$ meV. Optical conductivity $\sigma(\omega)$ is also gradually suppressed with decreasing temperature, and at 8 K, $\sigma(\omega)$ becomes almost zero below $\omega \sim 15$ meV. These results clearly indicate an optical gap ~ 15 meV developing below ≈ 80 K [57],

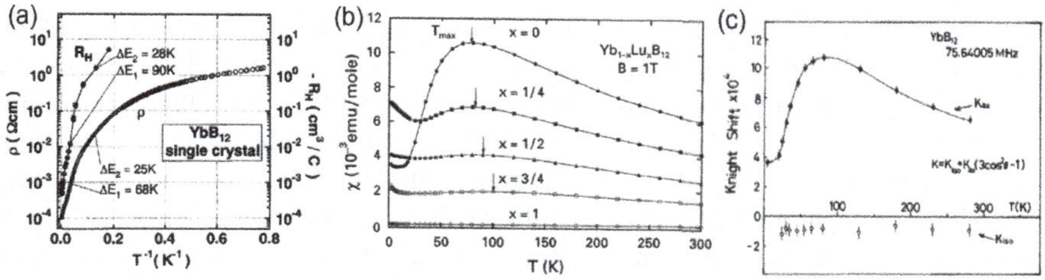

Fig. 1.12 **a** Arrhenius plot of the resistivity ρ and Hall coefficient R_{H} in a single crystal of YbB_{12}. Reprinted from [50], © Copyright 1998, with permission from Elsevier. **b** Temperature dependence of the susceptibility of $Yb_{1-x}Lu_xB_{12}$. Reprinted from [54], © Copyright 1999, with permission from Elsevier. **c** Temperature dependence of the ^{11}B NMR Knight shift. Reprinted from [55], © Copyright 1985, with permission from Elsevier

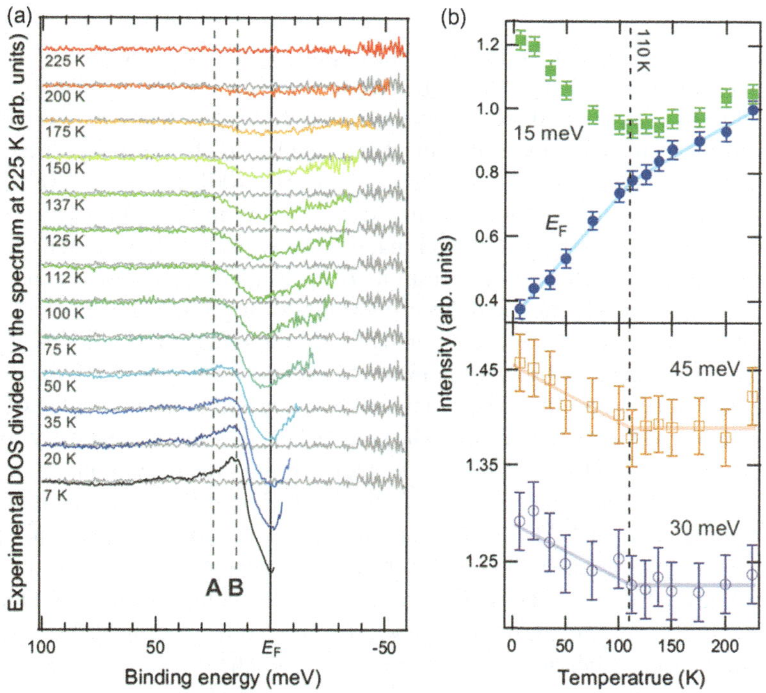

Fig. 1.13 Formation of hybridization gap in YbB_{12}. **a** DOS versus binding energy determined by laser PES measurements at various temperatures. **b** Temperature dependence of the spectrum intensity obtained from cuts at various binding energies in the data in (**b**). Reprinted figures with permission from [59] © 2015 by the American Physical Society

which is consistent with other experiments. Laser PES measurements also provide information on the DOS around E_F [59]. Figure 1.13b clearly shows the development of the multigap feature with temperature. Figure 1.13c also shows the temperature dependence of the spectral intensity with different binding energies obtained from the DOS data. The intensity at the inner-gap edge of 15 meV begins to develop below the coherent temperature of 110 K, and the DOS at E_F decreases owing to the gap formation. Although the coherent temperature is slightly higher than that determined from other experiments, the result strongly supports the opening of the hybridization gap at low temperature.

1.7.2 Field-Induced Insulator-Metal Transition

The hybridization gap can be suppressed by applying magnetic fields such that the Zeeman energy is sufficient to break the Kondo singlet: $g\mu_B B \sim \Delta$. The high magnetic field can completely destroy the gap so that the system turns metallic. Pulsed high-field measurements reveal this insulator-metal (I-M) transition with the critical field $H_{\text{I-M}}$ ranging from $\mu_0 H_{\text{I}-\text{M}} \simeq 45$–$47$ T ($\boldsymbol{H}||[100]$) to 55–59 T ($\boldsymbol{H}||[110]$) [60–63]. The signature for I-M transition can be found in the field dependence of resistivity [63], specific heat [62], and magnetization [61]. Reflecting the gap suppression, a negative MR is observed up to $H_{\text{I-M}}$. In the higher-field regime above $H_{\text{I-M}}$, Δ completely vanishes, and the system shows a negligibly small MR, indicating a field-induced metallic phase. Although not mentioned in this early paper [63], the oscillating behavior observed in an $x = 0$ sample around 40–55 T may be the signature of quantum oscillations, which will be discussed in more detail in Chap. 3. Corresponding to the phase transition, the specific heat C is also dramatically and discontinuously enhanced at the critical field. Figure 1.14a shows a B-T phase diagram with color map for the magnitude of C/T, which shows distinct phase boundary. The sudden change in C/T is also found in the C/T versus T^2 plots with different magnetic fields as shown in Fig. 1.14b. The intercept at $T^2 \rightarrow 0$ corresponds to Sommerfeld coefficient γ. Below $H_{\text{I-M}}$, γ is nearly zero within the experimental resolution. On the other hand, the finite large $\gamma \sim 60$ (mJ/molK^2) is observed above $H_{\text{I-M}}$, indicating that the DOS at E_F is induced in the metallic phase. The field-induced metallic phase with the large γ is termed the Kondo metal (KM) phase because the Kondo correlation does not break down at $H_{\text{I-M}}$ and robustly remains even in the metallic phase. The magnetization M also shows a kink anomaly at the critical field, and M shows a strong upturn in the KM regime. In the powder sample, a second transition at approximately 100 T is also observed [61]. While little is known about this phase because the excessively strong field limits the types of measurements we can conduct, we will examine recent magnetotransport experiments in the KM phase in Chap. 3.

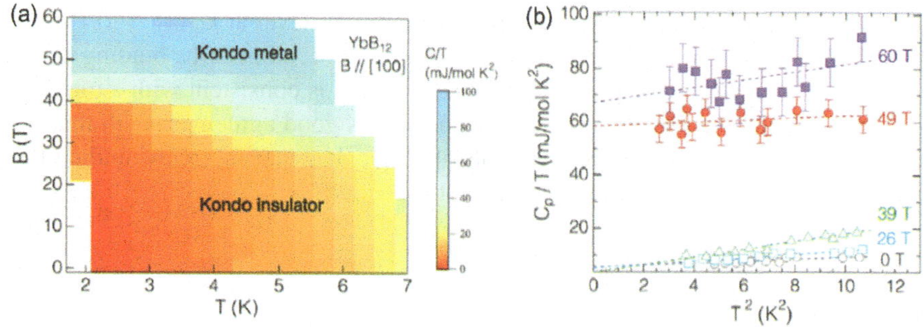

Fig. 1.14 **a** B-T mapping of specific heat (upper panel). **b** T^2 dependence of C/T with different magnetic fields. The intercept at $T^2 \to 0$ corresponds to the quasiparticle contribution. Reprinted figures with permission from [62] © 2018 by the American Physical Society

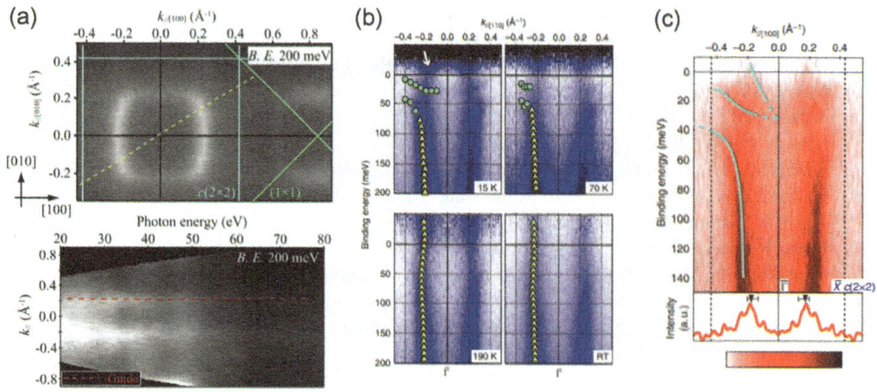

Fig. 1.15 ARPES data acquired from a clean (001) surface of YbB_{12}. **a** ARPES intensity map a the binding energy of 200 meV. **b** Band dispersion at different temperatures ranging from room temperature to 15 K. **c** ARPES data taken with a photon energy of 16.5 eV. The bottom panel shows the momentum distribution curve near $E_F = \pm 10$ meV. Reprinted from [64] by The Author licensed under CC BY 4.0

1.7.3 Topological Surface State

While SmB_6 has been intensively studied as the first candidate TKI, research on YbB_{12} is rather scarce. A recent ARPES measurement on a clean (001) surface revealed surface band dispersion, which is consistent with a topological SS [64]. As shown in Fig. 1.15a, the ARPES intensity at a biding energy of 200 meV shows distinct square constant-energy contours at 20 K. As indicated by the red dotted line, the photon energy does not affect the in-plane momentum of this band dispersion. This result indicates that the band does not host dispersion along k_z, demonstrating a 2D surface metallic state. Furthermore, this surface band shows hybridization with the flat f-band below T^*, as shown in Fig. 1.15b. The reconstructed band dispersion along the [100] direction taken with a photon energy of 16.5 eV is shown in

Fig. 1.15c. The surface band clearly crosses the Fermi level at $k_{||[100]} \sim 0.18$ Å^{-1}. This dispersion appears to connect to the f-band at the $\bar{\Gamma}$ point at the binding energy ~35 meV, leading to the degeneracy at TRIM, which is consistent with a topological SS. Although the surface band dispersion shows good agreement with that of a topological SS, the spin texture in the SS of YbB_{12} has been scarcely studied thus far, especially by means of transport properties.

References

1. Kondo J (1964) Prog Theor Phys 32:37
2. Wilson KG (1975) Rev Mod Phys 47:773
3. Andrei N, Furuya K, Lowenstein JH (1983) Rev Mod Phys 55:331
4. Haldane FDM (1978) Phys Rev Lett 40:416
5. Coleman P (2015) arXiv:1509.05769v1
6. Burdin S, Georges A, Grempel DR (2000) Phys Rev Lett 85(5):1048
7. Doniach S (1977) Phys B 91:231
8. Varma CM (2019) Rev Mod Phys 92:31001
9. Mott N (1974) Philos Mag 30(2):403
10. Dzero M, Xia J, Galitski V, Coleman P (2016) Annu Rev Condens Matter Phys 7:249
11. Kadowaki K, Woods SB (1986) Solid State Commun 58:507
12. Tsujii N, Kontani H, Yoshimura K (2005) Phys Rev Lett 94:057201
13. Franz R, Wiedemann G (1853) Ann Phys 165:497
14. Yamashita M, Nakata N, Senshu Y, Nagata M, Yamamoto HM, Kato R, Shibauchi T, Matsuda Y (2010) Science 328:1246
15. Murayama H, Sato Y, Xing XZ, Taniguchi T, Kasahara S, Kasahara Y, Yoshida M, Iwasa Y, Matsuda Y (2020) Phys Rev Res 2:013099
16. Grissonnanche G, Legros A, Badoux S, Lefrançois E, Zatko V, Lizaire M, Laliberté F, Gourgout A, Zhou JS, Pyon S, Takayama T, Takagi H, Ono S, Doiron-Leyraud N, Taillefer L (2019) Nature 571:376
17. Pfau H, Hartmann S, Stockert U, Sun P, Lausberg S, Brando M, Friedemann S, Krellner C, Geibel C, Wirth S, Kirchner S, Abrahams E, Si Q, Steglich F (2012) Nature 484:493
18. Machida Y, Tomokuni K, Izawa K, Lapertot G, Knebel G, Brison JP, Flouquet J (2013) Phys Rev Lett 110:236402
19. Mahajan R, Barkeshli M, Hartnoll SA (2013) Phys Rev B 88:125107
20. Dalum GARV, Mitchell AK, Fritz L (2020) Phys Rev B 102:041111(R)
21. Hasan MZ, Kane CL (2010) Rev Mod Phys 82:3045
22. Moore JE (2010) Nature 464:194
23. Fu L, Kane CL (2008) Phys Rev B 76:45302
24. He M, Sun H, He QL (2019) Front Phys 14:43401
25. Dzero M, Sun K, Galitski V, Coleman P (2010) Phys Rev Lett 104:106408
26. Menth A, Buehler E, Geballe TH (1969) Phys Rev Lett 22:295
27. Takimoto T (2011) J Phys Soc Jpn 80:123710
28. Lu F, Zhao J, Weng H, Fang Z, Dai X (2013) Phys Rev Lett 110:096401
29. Alexandrov V, Dzero M, Coleman P (2013) Phys Rev Lett 111:226403
30. Chen F, Shang C, Jin Z, Zhao D, Wu YP, Xiang ZJ, Xia ZC, Wang AF, Luo XG, Wu T, Chen XH (2015) Phys Rev B 91:205133
31. Eo YS, Rakoski A, Lucien J, Mihaliov D, Kurdak Ç, Rosa PFS, Fisk Z (2019) Proc Natl Acad Sci 116:12638
32. Röβler S, Jang TH, Kim DJ, Tjeng LH, Fisk Z, Steglich F, Wirth S (2014) Proc Natl Acad Sci 111:4798

33. Ruan W, Ye C, Guo M, Chen F, Chen X, Zhang GM, Wang Y (2014) Phys Rev Lett 112:136401
34. Zhang X, Butch NP, Syers P, Ziemak S, Greene RL, Paglione J (2013) Phys Rev X 3:011011
35. Neupane M, Alidoust N, Xu SY, Kondo T, Ishida Y, Kim DJ, Liu C, Belopolski I, Jo YJ, Chang TR, Jeng HT, Durakiewicz T, Balicas L, Lin H, Bansil A, Shin S, Fisk Z, Hasan MZ (2013) Nat Commun 4:2991
36. Xu N, Shi X, Biswas PK, Matt CE, Dhaka RS, Huang Y, Plumb NC, Radović M, Dil JH, Pomjakushina E, Conder K, Amato A, Salman Z, Paul DMK, Mesot J, Ding H, Shi M (2013) Phys Rev B 88:121102
37. Jiang J, Li S, Zhang T, Sun Z, Chen F, Ye ZR, Xu M, Ge QQ, Tan SY, Niu XH, Xia M, Xie BP, Li YF, Chen XH, Wen HH, Feng DL (2013) Nat Commun 4:3010
38. Hlawenka P, Siemensmeyer K, Weschke E, Varykhalov A, Sánchez-Barriga J, Shitsevalova NY, Dukhnenko AV, Filipov VB, Gabáni S, Flachbart K, Rader O, Rienks EDL (2018) Nat Commun. 9:517
39. Caldwell T, Reyes AP, Moulton WG, Kuhns PL, Hoch MJR, Schlottmann P, Fisk Z (2007) Phys Rev B 75:075106
40. Fuhrman WT, Chamorro JR, Alekseev PA, Mignot JM, Keller T, Rodriguez-Rivera JA, Qiu Y, Nikolić P, McQueen TM, Broholm CL (2018) Nat Commun 9:1539
41. Kim DJ, Xia J, Fisk Z (2014) Nat Mater 13:466
42. Syers P, Kim D, Fuhrer MS, Paglione J (2015) Phys Rev Lett 114:096601
43. Wolgast S, Kurdak Ç, Sun K, Allen JW, Kim DJ, Fisk Z (2013) Phys Rev B 88:180405
44. Thomas S, Kim DJ, Chung SB, Grant T, Fisk Z, Xia J (2016) Phys Rev B 94:205114
45. Kim J, Jang C, Wang X, Paglione J, Hong S, Lee J, Choi H, Kim D (2019) Phys Rev B 99:245148
46. Eo YS, Wolgast S, Rakoski A, Mihaliov D, Kang BY, Song MS, Cho BK, Hatnean MC, Balakrishnan G, Fisk Z, Saha SR, Wang X, Paglione J, Kurdak C (2020) Phys Rev B 101:155109
47. Xu N, Biswas PK, Dil JH, Dhaka RS, Landolt G, Muff S, Matt CE, Shi X, Plumb NC, Radović M, Pomjakushina E, Conder K, Amato A, Borisenko SV, Yu R, Weng HM, Fang Z, Dai X, Mesot J, Ding H, Shi M (2014) Nat Commun 5:4566
48. Moiseenko LL, Odintsov VV (1979) J Less-Common Metals 67:237
49. Kasaya M, Iga F, Negishi K, Nakai S, Kasuya T (1983) J Magn Magn Mater 31–34:437
50. Iga F, Shimizu N, Takabatake T (1998) J Magn Magn Mater 177–181:337
51. Weng H, Zhao J, Wang Z, Fang Z, Dai X (2014) Phys Rev Lett 112:016403
52. Teo JCY, Fu L, Kane CL (2008) Phys Rev B 78:045426
53. Legner M, Rüegg A, Sigrist M (2015) Phys Rev Lett 115:156405
54. Iga F, Hiura S, Klijn J, Shimizu N, Takabatake T, Ito M, Matsumoto Y, Masaki F, Suzuki T, Fujita T (1999) Phys B 259–261:312
55. Kasaya M, Iga F, Takigawa M, Kasuya T (1985) J Magn Magn Mater 47–48:429
56. Ikushima K, Kato Y, Takigawa M, Iga F, Hiura S, Takabatake T (2000) Phys B 281–282:274
57. Okamura H, Michizawa T, Nanba T, Kimura SI, Iga F, Takabatake T (2005) J Phys Soc Jap 74:1954
58. Susaki T, Sekiyama A, Kobayashi K, Mizokawa T, Fujimori A, Tsunekawa M, Muro T, Matsushita T, Suga S, Ishii H, Hanyu T, Kimura A, Namatame H, Taniguchi M, Miyahara T, Iga F, Kasaya M, Harima H (1996) Phys Rev Lett 77:4269
59. Okawa M, Ishida Y, Takahashi M, Shimada T, Iga F, Takabatake T, Saitoh T, Shin S (2015) Phys Rev B 92:161108(R)
60. Sugiyama K, Iga F, Kasaya M, Kasuya T, Date M (1988) J Phys Soc Jap 57:3946
61. Iga F, Suga K, Takeda K, Michimura S, Murakami K, Takabatake T, Kindo K (2010) J Phys Conf Ser 200:012064
62. Terashima TT, Matsuda YH, Kohama Y, Ikeda A, Kondo A, Kindo K, Iga F (2018) Phys Rev Lett 120:257206
63. Terashima TT, Ikeda A, Matsuda YH, Kondo A, Kindo K, Iga F (2017) J Phys Soc Jap 86:054710
64. Hagiwara K, Ohtsubo Y, Matsunami M, Ideta SI, Tanaka K, Miyazaki H, Rault JE, Le Fèvre P, Bertran F, Taleb-Ibrahimi A, Yukawa R, Kobayashi M, Horiba K, Kumigashira H, Sumida K, Okuda T, Iga F, Kimura SI (2016) Nat Commun 7:12690

Chapter 2
Quantum Oscillations in SmB_6

2.1 Introduction

A Fermi surface sets the boundary of occupied and unoccupied electron states in momentum space at zero temperature. It is well established that the presence of a Fermi surface is the definitive character of metals; hence, insulators (as well as semiconductors) do not have a Fermi surface. Most of the physical properties of a certain material are determined by its electrical ground state and how its electrons excite from that state. As electrons (fermions) follow the Pauli exclusion principle, only the electrons near the Fermi surface can participate in the excitations. This is why most of the physical properties of metals are determined by the geometry of the Fermi surface. Therefore, studying the geometry of the Fermi surface, which is also called Fermiology, is one of the most promising ways to understand a metal from a physical perspective.

Quantum oscillation (QO) is a phenomenon in which certain physical quantities, such as magnetization and resistivity, show periodic oscillations with respect to the reciprocal of external fields. QO is driven by the Landau quantization of conduction electrons in strong fields and reflects much information about the Fermi surface. Intuitively, this is because one can scan and map the electron states in momentum space by modulating the Landau level. Therefore, QOs have been employed as pivotal experimental tools in Fermiology. Recently, it has been reported that some TKIs such as SmB_6 and YbB_{12} show QOs, which is contrary to an expectation from the conventional theory of QOs. These observations have attracted great interest in the condensed matter physics community because these KIs are the first experimental counterexamples for insulators that may host Fermi surfaces. In this chapter, we review the recent observations of unconventional QOs in TKIs and the theories proposed to explain how they arise in insulators.

2.2 Quantum Oscillations

In classical mechanics, the motion of electrons under magnetic fields can be described as a cyclotron motion owing to the Lorentz force. This closed orbit is manifested

Y. Sato, *Quantum Oscillations and Charge-Neutral Fermions in Topological Kondo Insulator YbB_{12}*, Springer Theses,
https://doi.org/10.1007/978-981-16-5677-4_2

in the Landau quantization of energy levels (Landau levels) in the framework of quantum mechanics. The Schrödinger equation for a stationary charged particle in a 3D system is given by

$$\frac{1}{2m}(-i\hbar\nabla + e\mathbf{A})^2\psi = E\psi. \tag{2.1}$$

When a magnetic field B is applied parallel to the z-axis, the vector potential can be expressed as $\mathbf{A} = (0, Bx, 0)$. The Schrödinger equation can then be transformed into

$$\frac{\partial^2\tilde{\psi}}{\partial x^2} + \left[\frac{2mE}{\hbar^2} - k_z^2 - \left(\frac{eBx}{\hbar} - k_y\right)^2\right]\tilde{\psi} = 0, \tag{2.2}$$

where $\tilde{\psi} = \psi \exp[i(k_y y + k_z z)]$. By substituting $\xi_n = E - \hbar^2 k_z^2/2m$, $x' = x - \hbar k_y/eB$, and $\omega_c = eB/m$, we obtain the equation of motion for the harmonic oscillator:

$$\frac{\partial^2\tilde{\psi}}{\partial x^2} + \frac{2m}{\hbar^2}\left[\xi_n - \frac{1}{2}m\omega_c^2 x'^2\right]\tilde{\psi} = 0. \tag{2.3}$$

The energy eigenvalue is then simply given by

$$E = \xi_n + E(k_z) = \left(n + \frac{1}{2}\right)\hbar\omega_c + \frac{\hbar^2 k_z^2}{2m}. \tag{2.4}$$

Therefore, the energy levels for the motion within the xy-plane split into subbands with the energy difference $\Delta E = \hbar\omega_c$. Figure 2.1 shows schematics of the electron states of a 3D electron gas (a) without and (b) with an applied external field. Because the motion along the z-axis is not quantized, the energy levels form coaxial cylinders of Landau tubes along the z-axis.

The correspondence principle gives the orbit-area quantization condition [1], leading to the following relation between the cross section of the n-th cylinder A_n and n:

$$A_n = (n + \gamma)\frac{2\pi eB}{\hbar}. \tag{2.5}$$

As the field increases and energy splitting broadens, the outermost cylinders below the Fermi level are pushed out into the unoccupied states. This depopulation of the electrons near the Fermi level occurs periodically with respect to the change in the magnetic field, and its frequency F is given by

$$F = \frac{A(\theta)}{2\pi e\hbar}, \tag{2.6}$$

where $A(\theta)$ is the extreme cross section of the Fermi surface perpendicular to the field direction, as shown in Fig. 2.1c. Equation 2.6 is also known as Onsager's rule, according to which the frequency of the quantization is proportional to the cross

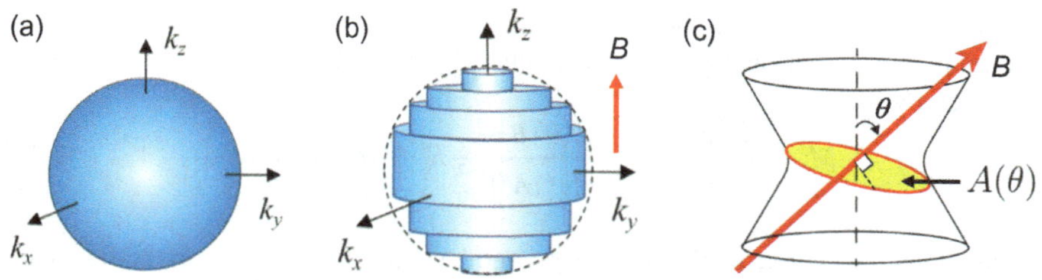

Fig. 2.1 Schematic of the spherical Fermi surface of a 3D free electron gas **a** without external fields and **b** under a magnetic field along the z-direction. **c** Extreme cross section $S(\theta)$ of the Fermi surface perpendicular to the magnetic field tilted by θ from the z-axis. © 2021 The author.

section of the Fermi surface. Consequently, the thermodynamic potential shows sets of oscillations as a function of the reciprocal of the field; consequently, some physical properties such as the thermodynamic and transport quantities of metals also show oscillations. The QOs in magnetization are called the de Haas–van Alphen (dHvA) effect, while the QOs in resistivity are referred to as the Shubnikov–de Haas (SdH) effect.

It is worth noting that Eq. 2.6 holds even if A_n is B-dependent. For the B-dependent cross section $A(B_n)$, Eq. 2.5 can be expressed as

$$A(B_n) = (n + \gamma)\frac{2\pi e B_n}{\hbar}. \tag{2.7}$$

A set of equations for n and n+1 can be combined to produce

$$\frac{A(B_{n+1})}{B_{n+1}} - \frac{A(B_n)}{B_n} = \frac{2\pi e}{\hbar}. \tag{2.8}$$

Moreover, F is field-dependent, and the quantization condition should be satisfied when the field is equal to $F(B_n)$, i.e., $F(B_n)/B_n = n + \gamma$, which yields

$$\frac{F(B_{n+1})}{B_{n+1}} - \frac{F(B_n)}{B_n} = 1. \tag{2.9}$$

Therefore, Eqs. 2.8 and 2.9 demonstrate that Onsager's rule still holds

$$F(B) = \frac{\hbar}{2\pi e} A(B). \tag{2.10}$$

The QOs are formulated as the well-known Lifshitz–Kosevich (LK) formula [2, 3], and the oscillatory part of magnetization is given by

$$M_{\mathrm{osc}} = -\sum_{r=1}^{\infty} \frac{1}{r^{3/2}} M_r \sin\left[2\pi r\left(\frac{F}{B} - \frac{1}{2} \pm \frac{\pi}{4}\right)\right], \tag{2.11}$$

where the $+$ and $-$ signs in the phase shift of $\pm\pi/4$ correspond to the minimum and maximum cross-sectional area, respectively. The amplitudes of the oscillations M_r are given by

$$M_r = \left(\frac{e}{2\pi\hbar}\right)^{3/2} \frac{A(\theta)B^{1/2}}{\pi^2 m^* |A''|_{\text{extr}}^{1/2}} R_T(r)R_{\text{D}}(r)R_S(r), \tag{2.12}$$

where m^* is the cyclotron effective mass, and $|A''|_{\text{extr}} = (\partial^2 S/\partial p_B^2)_{\text{extr}}$ is the curvature of the Fermi surface around the cross section along the B direction. Equation 2.12 also contains three types of damping factors, namely temperature, Dingle, and spin damping, which are denoted as $R_T(r)$, $R_{\text{D}}(r)$, and $R_S(r)$, respectively. These three factors arise from the broadening of the Landau levels driven by the finite thermal energy in the Fermi–Dirac distribution function, impurity scattering, and Zeeman splitting of opposite spin bands, respectively. $R_T(r)$ can be expressed as

$$R_T(r) = \frac{2\pi^2 r k_{\text{B}} T/(\hbar\omega_{\text{c}})}{\sinh[2\pi^2 r k_{\text{B}} T/(\hbar\omega_{\text{c}})]} = \frac{Kr\mu T/B}{\sinh(Kr\mu T/B)}, \tag{2.13}$$

where $\mu = m^*/m_0$ and $K \equiv 2\pi^2 k_{\text{B}} m_0/(\hbar e) \sim 14.7$ T/K. Therefore, by utilizing μ as a fitting parameter, one can estimate the effective mass of electrons from the temperature dependence of the oscillatory amplitudes. The Dingle damping factor is given by

$$R_{\text{D}}(r) = \exp\left(-\frac{\pi r}{\omega_{\text{c}}\tau}\right) = \exp(-B_{\text{c}}/B) = \exp(-Kr\mu T_{\text{D}}/B), \tag{2.14}$$

where $T_{\text{D}} = \hbar/(2\pi k_{\text{B}}\tau)$ is the Dingle factor related to the impurity scattering energy. The characteristic field found in the numerator of the exponent $B_{\text{c}} = Kr\mu T_{\text{D}}$ is the characteristic field above which QOs are visible. The field B_c can also be expressed as the reciprocal of the mobility $e\tau/m^*$. By utilizing T_{D} as a fitting parameter again, one can estimate the scattering time τ from the field dependence of the oscillatory amplitudes. Finally, the spin-damping factor is written as

$$R_S(r) = \cos\left(\frac{\pi}{2} r g \mu\right), \tag{2.15}$$

which contains the g-factor. As shown above, the measurements of QOs provide rich information on the geometry of the Fermi surface, effective mass, scattering time, and g-factor of metals. According to Eqs. 2.13 and 2.14, the oscillation amplitude becomes visible when $\hbar\omega_{\text{c}} > k_{\text{B}}T$ and $\hbar\omega_{\text{c}} > \hbar/\tau$. The first condition is rewritten as $B/T > k_{\text{B}} m^*/\hbar e$, suggesting that sufficiently large magnetic fields and low temperatures are required to experimentally resolve QOs. Moreover, when the cyclotron mass m^* is low, QOs are relatively easy to detect. The second condition from the Dingle factor leads to $\tau > 1/\omega_{\text{c}}$, implying that an electron must form a closed cyclotron orbit before it is scattered. This condition necessitates a clean crystal with a long scattering time.

2.3 Results Reported by the Michigan Group

The first observations of QOs in the KI SmB_6 were reported by Li et al. [4] from the University of Michigan. They synthesized single crystals of SmB_6 by using the aluminum flux method and measured the magnetic torque $\boldsymbol{\tau}$ by using the capacitance cantilever method. As $\boldsymbol{\tau} = \boldsymbol{M} \times \boldsymbol{H}$, the oscillations in $\boldsymbol{\tau}$ are a direct consequence of the dHvA effect. The clear oscillations in the magnetization are observed at high fields [4]. By performing a fast Fourier transformation (FFT), one can separate multiple components of oscillations with different frequencies F. It is reported that one of the characteristic frequencies, F^{β}, as a function of the field angle ϕ shows four-fold rotational symmetry, which is expected from the cubic crystalline structure of SmB_6. Most importantly, F^{β} can be perfectly fitted by

$$F = \frac{F_0}{\cos(\phi - \pi/4 - n\pi/2)}, \tag{2.16}$$

which is the angular dependence of the cross section expected from the cylindrical Fermi surface of 2D electrons. This field dependence strongly suggests that the dHvA effect originates from the 2D electronic state, which is most likely to have originated from the topological SS. The temperature dependence of the amplitude of the oscillations is also discussed. As the oscillations contain three different frequencies and, thereby, different Fermi pockets α, β, and γ, one must separate each oscillatory component by performing FFT. It is reported that the normalized amplitudes follow the LK formula given by Eq. 2.13, which extracts the effective mass $m^* = 0.12$–$0.19 m_0$. The observed LK behavior indicates that the quasiparticles participating in the dHvA effect follow the Fermi–Dirac statistics, and hence, they are fermions. The low effective mass of roughly 0.1 times that of a free electron is rather surprising, as the strong correlations in KI tend to yield heavy fermions with a large m^*.

2.4 Results Reported by the Cambridge Group

Soon after the first report by the Michigan group, Tan et al. [5] from Cambridge University also reported the dHvA effect in SmB_6. Although they found the same phenomenon in the same compounds, they presented some surprising results, which were not observed in the first report. While they employed the same types of experimental techniques, the single crystals of SmB_6 used in their study were prepared using the floating zone method. In addition to the low-frequency oscillations around $F \sim 10^2$–10^3 T, which were also observed by the Michigan group, they found other oscillations with higher frequencies $F \sim 10^3$–10^4 T. It was pointed out that the observed higher frequency in SmB_6 and its angular dependence is akin to that of an isostructual metallic compound LaB_6, which implies that these two materials have similar Fermi surfaces. As one can regard LaB_6 as SmB_6 without hybridization owing to the absence of f-electrons, its Fermi surfaces are nearly spherical large pockets centered upon the X points. This similarity of the Fermi surfaces is quite surprising because such a large 3D Fermi surface rules out the possibility that the dHvA effect with the higher frequency originates from the SS. Therefore, the Cambridge group's results suggest that SmB_6 somehow possesses an unconventional 3D bulk Fermi surface, although it is a charge insulator. As this was the first experimentally observed insulator that shows QOs, the oscillations are occasionally referred to as unconventional QOs.

Another surprising feature of this unconventional dHvA effect is its non-LK behavior in the temperature damping factor. A good agreement with the LK curve is reported around 1–15 K, yielding the light effective mass of $m^* \approx 0.18m_0$. This m^* is rather close to that observed by the Michigan group. However, the amplitude strongly deviates from the LK fit below ~1 K, showing a non-monotonic, sharp upturn toward low temperatures. This unusual temperature dependence of amplitude may indicate that the quasiparticles do not follow the Fermi–Dirac distribution, or that multiple types of quasiparticles with different m^* are involved in the dHvA signals. The theoretical approach for this non-LK behavior will be discussed in Sect. 2.6.

2.5 Problems Related to the QOs in SmB_6

Although the observations of the dHvA effect provide convincing evidence for the Fermi surfaces of SmB_6, which may originate from the topological SS or exotic 3D insulating phase, several problems remain. The first problem is why the QOs in resistivity (SdH effects) have never observed so far. The MR of the flux-grown sample was also reported by the Michigan group [4]. Although the dHvA effects are clearly resolved below T ~25 K and above B ~5 T, no evidence of SdH oscillations was detected, even at the lowest temperature of T ~350 mK and fields of up to ~45 T. Given the low m^* and high mobility of the quasiparticles estimated from the dHvA signals, the absence of the SdH signal in such conditions is puzzling. A possible explanation for this issue is that the scattering probability is not significantly affected by the Landau quantization, leading to less pronounced SdH signals, while the dHvA effect is a direct consequence of the oscillations in the thermodynamic potential. However, this idea has not been quantitatively justified, and the lack of the SdH signal has been under debate.

The second issue is whether the dHvA effects are intrinsic properties of SmB_6. As the difference between the results of the Michigan group and Cambridge group lies in the method of crystal growth, it has been suggested that the qualitative difference might have originated from the sample quality or an extrinsic effect induced by the impurity of the aluminum flux. From this perspective, torque measurements on SmB_6 were re-examined by Thomas et al. [6] from Los Alamos National Laboratory. To verify the effect of the aluminum flux more clearly, they prepared crystals with an intentionally large amount of aluminum flux, as shown in Fig. 2.2a. The torque signal with respect to the field is shown by the blue curve in Fig. 2.2b. The crystal

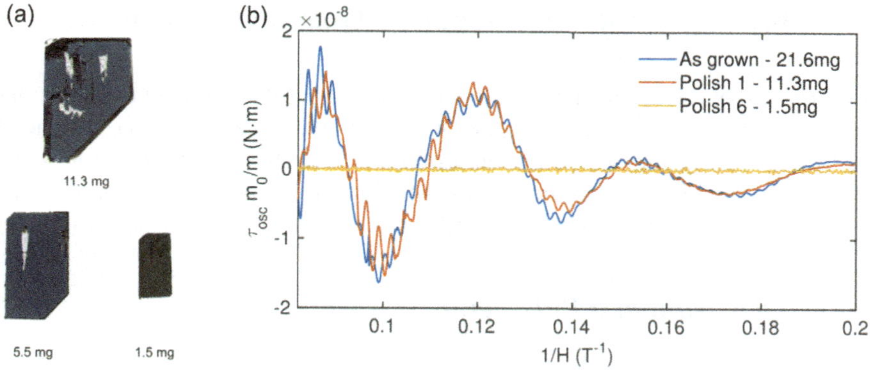

Fig. 2.2 **a** SmB_6 crystal with intentionally embedded aluminum impurities. The sample was polished into small pieces with a low concentration of aluminum. **b** dHvA data observed in the samples shown in (**a**). Reprinted figures with permission from [6] © 2019 by the American Physical Society.

shows a clear dHvA effect, and the angular dependence of this frequency and the temperature dependence of the amplitude are quite similar to those reported by the Michigan group. Furthermore, on polishing the as-grown crystal to remove the flux, the oscillations in torque decrease and finally vanish within the experimental resolution. The results suggest that the 2D features of the dHvA effect originate from the extrinsic metallic aluminum phase and are not intrinsic to SmB_6.

On the other hand, the Cambridge group recently reported another experiment [7], where they argue that the dHvA effect is an intrinsic property of the insulating bulk of SmB_6, as reported earlier. They prepared an extremely clean single crystal of SmB_6 by using the floating zone method. The crystals show a lower concentration of impurities, better thermal conductivity, and an inverse residual resistivity ratio higher than those of previously investigated samples by over an order of magnitude, demonstrating that the new sample is of much higher quality. In the newly prepared crystals, they reproduced dHvA signals similar to those in the previous report [5]. The oscillation amplitudes are comparable to that of the paramagnetic response, indicating that the dHvA effects originate from a major portion of the sample, i.e., the bulk. Moreover, they also show the dHvA signals of a single crystal of elemental aluminum, pointing out that their frequencies are essentially different from that observed in SmB_6.

As seen thus far, there is still no consensus on whether the QOs have a surface or bulk origin; it is also unclear whether the dHvA effect is intrinsic to SmB_6. It is, thus, essential to address these issues for understanding the nature of unconventional QOs in KIs, and further investigation on other materials is necessary.

2.6 Theories for the Unconventional Quantum Oscillations

After the discovery of the dHvA effects in SmB_6, a substantial number of theories have been proposed to explain this phenomenon. In this section, we review some of the theories to examine how QOs can arise from TKIs and how these theories can be verified experimentally. Some are based on the intrinsic properties of narrow-gap insulators, while some assume certain types of charge-neutral quasiparticles.

2.6.1 QOs Without a Bulk Fermi Surface

It was pointed out that the gap amplitude of KIs shows oscillatory narrowing under external fields owing to the inverted band structure resulting from the c-f hybridization [8]. In a simple metal with a parabolic conduction band, the Landau tubes periodically swell out and pass through the chemical potential μ as the field increases, leading to the periodic modulation of the low-energy density of states (LEDOS), as shown in Fig. 2.3a. In KIs, on the other hand, the Landau tubes first approach μ but are reflected at the band edge; consequently, they move away from μ (Fig. 2.3b). In accordance with this gap narrowing, the LEDOS also oscillates. In a narrow-gap insulator, thermally activated quasiparticles show periodic oscillations in the thermodynamic potential, leading to the magnetic QOs. A key ingredient for this theory is only the inverted band structure, which is preferred in KIs; therefore, any narrow-gap insulators with band inversion would show QOs irrespective of their topological nature. It is worth noting that this thermal activation leads to a exponential decrease in the oscillatory amplitude as approaching to the absolute zero temperature.

There are also several theories based on the topological SS as a source of the QOs [9]. In this scheme, the non-LK behavior observed in SmB_6 can be explained by the Kondo breakdown of the SS. The band structures of SmB_6 are calculated by changing the parameter $\langle b_s \rangle^2/\langle b \rangle^2$, which indicates the amount of hybridization

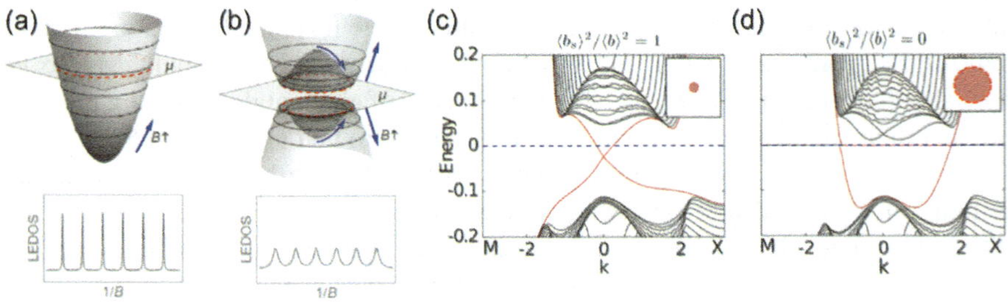

Fig. 2.3 Schematic of the band structure and the corresponding LEDOS in **a** metals and **b** KIs. The blue arrow indicates the direction to which the Landau level moves as the field increases. Reprinted figures with permission from [8] © 2016 by the American Physical Society. **c**, **d** Calculated surface (red lines) and bulk (black lines) band structures of KI with different hybridization strengths on the SS. Reprinted figures with permission from [9] © 2016 by the American Physical Society.

between the local moments on the surface and the topological SS. Figure 2.3c and d shows the resulting band structure of the bulk and the SS with black and red curves, respectively, and their insets depict the Fermi surface of the SS. For $\langle b_s \rangle^2/\langle b \rangle^2 = 1$, the SS is completely hybridized to the local moments on the surface, leading to surface electrons with a large m^* forming a small Fermi pocket. With decreasing temperature, $\langle b_s \rangle^2/\langle b \rangle^2$ decreases, and the surface electrons decouple to the local moment, resulting in the small m^* and large Fermi surface. This Kondo breakdown on the surface possibly explains the non-LK behavior through the dramatic change in m^*.

An alternative approach is based on non-Hermitian Landau level problem of impurity-induced in-gap states of narrow insulators [10, 11]. The electron scattering between different Landau levels can be described by the non-Hermitian quasiparticle Hamiltonian. This scattering process is important to realize the in-gap state, which is responsible for the QOs. Contrary to the gap-narrowing scenario, the amplitude of QOs remains finite even at the absolute zero temperature in this theory.

The importance of the strong correlation effect to the SS has also been studied [12]. The interplay between correlations and topology successfully explains both the dHvA and SdH effect, the amplitudes of which are enhanced on increasing the correlation strength.

2.6.2 *Neutral Quasiparticles as a Source of QOs*

The theories we have discussed so far do not require any actual bulk Fermi surfaces but demonstrate that some insulators still can exhibit QOs. A more straightforward idea, in some ways, is that a certain type of quasiparticles forms bulk Fermi surfaces and can experience the Landau quantization in magnetic fields, resulting in the unconventional QOs. However, such quasiparticles cannot be conventional electrons because they are not directly responsible for the charge transport, given the activation behavior in the resistivity of KIs. In other words, they must be charge-neutral quasiparticles. Some theories are, in fact, based on this idea, and several types of charge-neutral quasiparticles have been proposed.

Knolle and Cooper proposed that excitons, which are bound states of an electron and an hole, can be realized in SmB_6 and may be the source of the QOs [13]. In this scenario, they showed that KIs are susceptible to the formation of excitons owing to the ring-shaped dispersion of the hybridized bands. In a certain range of parameters, the dispersion of the excitons can be considered gapless, which results in the QOs. Despite the charge neutrality of excitons, they are responsible for the thermal excitations, leading to the finite specific heat and thermal conductivity. Because of their bosonic nature, the calculated temperature dependence of specific heat and thermal conductivity shows non-monotonic behavior, as shown in Fig. 2.4a–b. While this theory can explain the dHvA effect and its non-LK temperature damping, it is not unclear how SdH effects can be described within this theoretical framework.

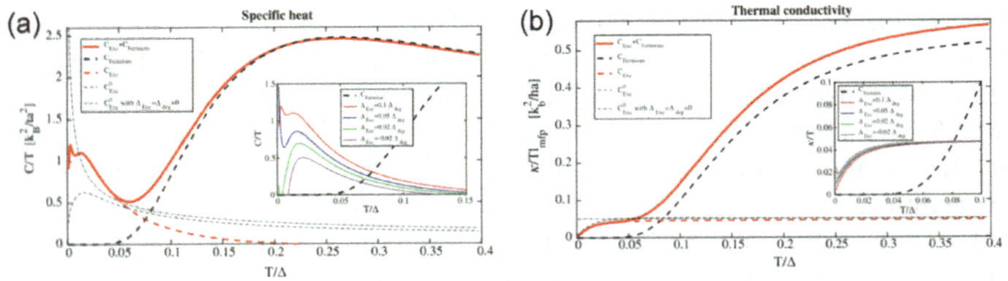

Fig. 2.4 Calculated temperature dependence of the **a** specific heat and **b** thermal conductivity based on the exciton model. Reprinted figures with permission from [13] © 2017 by the American Physical Society.

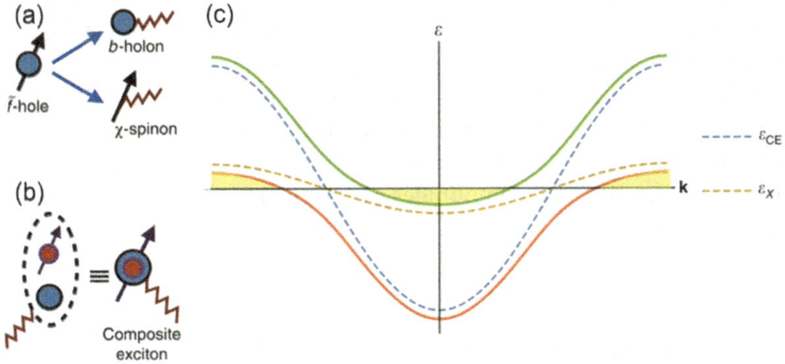

Fig. 2.5 Proposed Fermi surfaces of neutral fermions of composite excitons. **a** Schematic of the fractionalization of the f-hole. **b** Formation of the composite exciton between a conduction electron and holon. **c** Hybridized band of the neutral fermion of an exciton and a spinon. Reprinted from [14] by The Author licensed under CC BY 4.0.

Chowdhury et al. also developed a theory based on excitons [14]. First, following the slave boson representation $f = b\chi_s$, the f-holes f are fractionalized into spinless bosons b (holons) with charge $-e$ and neutral fermions χ_s (spinons) with spin s (Fig. 2.5a–b). The conduction d-electrons d then form binding states with holons b in the form $\Psi = bd$ owing to the strong attractive interaction. Here, Ψ represents the composite fermionic excitons that follow Fermi–Dirac statistics. As these composite excitons and spinons form neutral Fermi surfaces of hybridized bands, as shown in Fig. 2.5c, the low-energy excitations become similar to that of ordinal metals, yielding T-linear terms in the specific heat $C \sim \gamma T$, thermal conductivity $\kappa \sim \kappa_0 T$, and NMR spin-lattice relaxation rate, $1/T_1 \sim T$. This theoretical view is similar to the spinon Fermi surface proposed in quantum spin liquids [15], although QOs have never been experimentally realized in this class of materials. QOs can also occur when these neutral fermions couple to external fields through the internal gauge degree of freedom in holons. This coupling is also expected to be manifested as a sizable finite thermal Hall conductivity κ_{xy}, as the neutral fermions experience the Lorentz force in fields.

Majorana fermions have also been proposed [16–18] as a source of the unconventional QOs. Historically, they were first introduced in the context of particle physics, and recently, they have been intensely debated as possible emergent quasiparticles in certain quantum materials such as the Kitaev model [19, 20]. It is interesting to note that the introduction of the Majorana operator in KI was first prompted by Coleman et al. in 1993 before the discovery of TIs and the unconventional QOs in KIs [21]. Majorana fermions are particles whose antiparticles are themselves ($\gamma = \gamma^\dagger$), and therefore, they are charge neutral. They can be described as fractionalized electrons and may form a neutral Fermi surface, as shown in Fig. 2.5d. The QOs can be realized through the Landau quantization of original electrons. However, no experiments have provided direct evidence of Majorana fermions in KIs.

We have seen that several types of neutral quasiparticles have been proposed as possible candidates for the origin of the QOs. One way to test the presence of such neutral particles is to detect the neutral Fermi surface by measuring other physical quantities such as thermal conductivity. The response with respect to an applied magnetic field may also provide pivotal information that allows us to identify the microscopic origin of the excitations.

References

1. Onsager L (1952) Philos Mag 43:1006
2. Lifshitz I, Kosevich A (1956) Sov. Phys JETP 2:636
3. Kartsovnik MV (2004) Chem Rev 104:5737
4. Li G, Xiang Z, Yu F, Asaba T, Lawson B, Cai P, Tinsman C, Berkley A, Wolgast S, Eo YS, Kim DJ, Kurdak C, Allen JW, Sun K, Chen XH, Wang YY, Fisk Z Li L (2014) Science 346:1208
5. Tan BS, Hsu YT, Zeng B, Ciomaga Hatnean M, Harrison N, Zhu Z, Hartstein M, Kiourlappou M, Srivastava A, Johannes MD, Murphy TP, Park JH, Balicas L, Lonzarich GG, Balakrishnan G, Sebastian SE (2015) Science 349:287
6. Thomas SM, Ding X, Ronning F, Zapf V, Thompson JD, Fisk Z, Xia J, Rosa PFS (2019) Phys Rev Lett 122:166401
7. Hartstein M, Liu H, Hsu Y-T, Tan BS, Hatnean MC, Balakrishnan G, Sebastian SE (2020) Science 23:101632
8. Zhang L, Song XY, Wang F (2016) Phys Rev Lett 116:046404
9. Erten O, Ghaemi P, Coleman P (2016) Phys Rev Lett 116:046403
10. Shen H, Fu L (2018) Phys Rev Lett 121:026403
11. Yoshida T, Peters R, Kawakami N (2018) Phys Rev B 98:035141
12. Peters R, Yoshida T, Kawakami N (2019) Phys Rev B 100:085124
13. Knolle J, Cooper NR (2017) Phys Rev Lett 118:096604
14. Chowdhury D, Sodemann I, Senthil T (2018) Nat Commun 9:1766
15. Balents L (2010) Nature 464:199
16. Baskaran G (2015) arXiv:1507.03477
17. Erten O, Chang PY, Coleman P, Tsvelik AM (2017) Phys Rev Lett 119:057603
18. Varma CM (2020) Phys Rev B 102:155145
19. Kitaev A (2006) Ann Phys 321:2
20. Kasahara Y, Ohnishi T, Mizukami Y, Tanaka O, Ma S, Sugii K, Kurita N, Tanaka H, Nasu J, Motome Y, Shibauchi T, Matsuda Y (2018) Nature 559:227
21. Coleman P, Miranda E, Tsvelik A (1993) Physica B 186–188:362

Chapter 3
Unconventional Quantum Oscillations in YbB_{12}

3.1 Introduction

The observations of unconventional QOs in SmB_6 stimulated substantial efforts to understand its nature, both theoretically and experimentally. The fundamental question is whether they are intrinsic properties commonly observable among (topological) KIs or a specific feature inherent in SmB_6. A comparison to other similar materials, thus, may provide important information to solve this problem. In this chapter, we review observations of QOs in other TKI YbB_{12} [1, 2] and discuss possible scenario, where charge-neutral fermions are the source of the QOs. High-field elcectric properties such as the transport, magnetic torque, and penetration depth were investigated in both insulating and field-induced metallic phases by employing a series ofc experimental techniques.

3.2 Sample Characterization

YbB_{12} samples studied and presented in this chapter are characterized as follows. First, high-quality single crystals of YbB_{12} were grown and provided by Professor Fumitoshi Iga at Ibaraki University, Japan. The crystals were synthesized using the traveling-solvent floating zone method [3]. A series of measurements were performed for three different samples, which are labeled #1, #2, and #3. #1 and #2 were cut off from the same growth batch, while #3 was taken from another batch. To estimate the sample quality, synchrotron X-ray powder diffraction measurements were also conducted at the BL02B2 beamline at the SPring-8 facility, Japan. Figures 3.1a, b display the diffraction pattern. Fine peaks attributed to the crystalline structure of YbB_{12} were clearly observed. As impurities, tiny amounts of Al_2O_3 and YbB_6 were found. Al_2O_3 contamination possibly occurred during the grinding process for making the powder from the bulk crystal, while YbB_6, which is a correlated insulator, is the only impurity phase found in the samples. Peaks corresponding to other metallic YbB_x compounds such as YbB_2 and YbB_4 could not be resolved with the experimental resolution, and the maximum volume can be estimated to be less

Y. Sato, *Quantum Oscillations and Charge-Neutral Fermions in Topological Kondo Insulator YbB_{12}*, Springer Theses,
https://doi.org/10.1007/978-981-16-5677-4_3

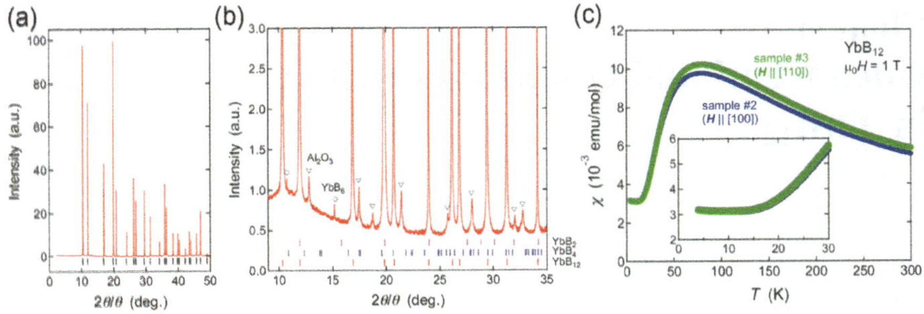

Fig. 3.1 Sample characterizations. **a**, **b** Synchrotron X-ray powder diffraction pattern of the YbB_{12} samples. The bottom tick marks in (**a**) represent the peak positions of YbB_{12}, while the circles and triangles in (**b**) indicate the peaks of YbB_6 and Al_2O_3, respectively. **c** Temperature dependence of the susceptibility χ in #2 and #3. The inset shows an expanded view of the region below 30 K. © 2021 The author.

than 10 ppm. These diffraction results indicate that the samples contain an extremely low concentration of metallic impurities. The magnetic susceptibilities of crystal #2 and #3 were measured using MPMS Quantum Design, and the resulting temperature dependence is shown in Fig. 3.1c. An excellent agreement with previously reported data [4] was found: χ shows a Curie–Weiss-like response at high temperature as well as a broad peak around 70 K, and it finally shows saturation at low temperature. However, there is a minor difference in that the absolute value below 10 K is slightly smaller than that in the previous report. This suggests that Kondo screening is a little stronger in the present sample, which may indicate a better sample quality.

The higher sample quality also results in a slight variation in the gap amplitude Δ among the different growth batches. Figure 3.2a displays the temperature evolution of the resistivity ρ of all the three samples. At 0.1 K, ρ is 4–5 orders of magnitude larger than that at room temperature. Although ρ is not sample-dependent around room temperature, it shows significant variation with decreasing temperatures. The resistivity plateau, which we also discussed in Chap. 1, appears to be due to surface conduction as the residual resistivity is roughly proportional to the sample thickness. An Arrhenius plot, $\ln\rho$ versus $1/T$, is used to evaluate Δ for all crystals. A two-gap behavior was found, which is consistent with a previous transport measurement [3]. The small gap developing below ≈ 20 K shows a slight sample dependence, and the linear fittings over the temperature range of $6\,\mathrm{K} < T < 12.5\,\mathrm{K}$ yield a gap of 4.7 meV for #1 and #2 and of 4.0 meV for #3. The present results suggest that the sample quality of the first batch (#1 and #2) is slightly higher than that of the second one (#3), as the large activation energy, in general, implies a low impurity concentration and high homogeneity. The larger gap developing in the temperature range of $20\,\mathrm{K} < T < 300\,\mathrm{K}$ is approximately 11 meV, and this gap does not show significant sample dependence. We note that the larger gap is almost twice larger than that reported in another paper [5].

It is also worth noting that the transport data do not reveal any metallic behavior below the I-M transition critical field $\mu_0 H_{\text{I-M}} \sim 46$ T. Figure 3.2c shows the temper-

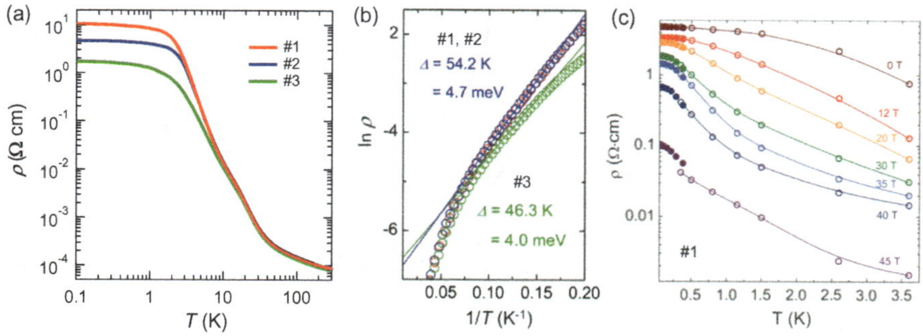

Fig. 3.2 **a** Temperature dependence of resistivity in three samples. **b** Arrhenius plot above 5 K. The solid lines indicate the fitting results, and the activation energy Δ is extracted from their slopes. **c** Temperature dependence of resistivity in various fields taken using pulsed fields. The open and solid circles represent the data taken in a ^{3}He cryostat at $\phi = 7.4°$ and in a dilution fridge at $\phi = 8.5°$, respectively. The solid lines are guides to the eye. © 2021 The author

ature dependence of the resistivity with different magnetic fields of up to 45 T. Even at 45 T, ρ shows an increasing trend with decreasing temperature, indicating that the activation behavior survives and the samples are purely insulating.

3.3 QOs in the Insulating Phase of YbB_{12}

Figure 3.3a shows the field dependence of the magnetic torque τ in #1 with different ϕ of up to 45 T. Here, ϕ denotes the angle of the magnetic field direction from the [100] crystal axis. τ shows a step-like anomaly around 20 T, which is attributed to the meta-magnetic transition. More importantly, τ shows distinct oscillations around ~38–45 T. It is remarkable that the dHvA effect occurs in the insulating phase of YbB_{12}, similar to SmB_6. The inset shows the FFT frequency of the dHvA oscillations in the field range of 38.5–45 T. The major peak of $F = 720$ T and its higher harmonics are clearly resolved by the calculation.

Next, high-field MR data in the three samples are given in Fig. 3.3b. Remarkably, the samples show clear oscillations well below the critical field $H_{\text{I-M}}$, revealing that the SdH effect can be observed in the insulating phase of YbB_{12}, unlike SmB_6. The negative slope of the background MR, which is the hallmark of field-induced gap suppression, can be subtracted by polynomial fitting to produce the oscillatory component of the MR $\Delta\rho$, as shown in Fig. 3.3c. Distinct oscillations were resolved especially in #1 and #2, but #3 does not show significant periodic modulations. The fact that #1 and #2 show peaks at the same intervals with respect to $1/H$ strongly indicates that the oscillation is indeed the SdH effect. Moreover, the SdH oscillations in #1 and #2 become visible under almost the same field, implying that the scattering rates in these two samples are almost identical. On the other hand, because #3 has a smaller Δ than #1 and #2, the scattering rate in #3 should be lower. The absence of the

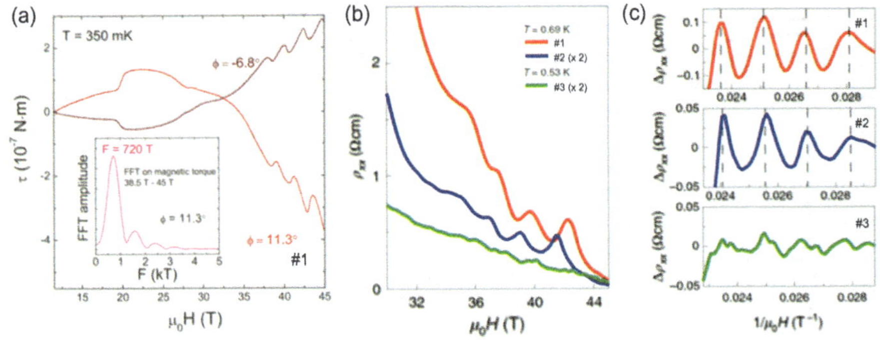

Fig. 3.3 Quantum oscillations observed in the insulating phase of YbB_{12}. **a** dHvA effect proved by the magnetic torque. The inset shows the oscillation frequency calculated via FFT. **b** High-field magnetoresistance of three samples. #1 and #2 show distinct SdH oscillations around 36–44 T. **c** Oscillatory component of the magnetoresistance as plotted against $1/\mu_0 H$ for all samples, obtained from the data in (**b**). © 2021 The author

SdH effect in #3 is consistent with this view, but it also indicates that crystals with stronger insulating characters show more pronounced QOs, which is inconsistent with the conventional theory of QOs. The current observations imply that the QOs in the insulating phase are an inherent feature of YbB_{12}, as metallic impurities, if they existed, would have behaved in a manner opposite to the observations.

As discussed in Sect. 2.6, it is important to check whether the quasiparticles forming the Fermi surface follow the Fermi–Dirac distribution law. The temperature dependence of the normalized oscillation amplitude for both the dHvA and SdH effects is displayed in Fig. 3.4a–c. The solid curves are the fitting results obtained using the LK formula (Eq. 2.13). There is an excellent agreement with the LK curves for the data of both oscillations over the different ranges of fields and angles, implying that the quasiparticles are fermions. The fittings yield an effective mass ratio of m^*/m_0 ~6.7 from the dHvA effect and ~15 from the SdH effect. The different effective masses obtained from the dHvA and SdH effects may imply that they originate from different bands, although there remains the question why the band with the smaller mass detected from the dHvA effect does not contribute to the SdH signal. The observed effective mass is larger than that of a bare electron, which is contrary to the case of the small mass observed from the dHvA effect in SmB_6. This larger mass of the quasiparticles indicates that they result from the strong correlation in the Kondo lattice. The successful FFT calculation of the oscillations and the good agreement with the LK description confirm that the observed high-field features are QOs, rather than a series of field-induced Lifshitz transitions.

The angular dependence of F, which is calculated using FFT and Landau level fitting for each type of QOs, is displayed in Fig. 3.5a. The open circles represent the data obtained from the dHvA effect, whereas the open triangles represent data extracted from the SdH oscillations. The angular dependences of F obtained from the dHvA and SdH effects are quite different, indicating that they originate from

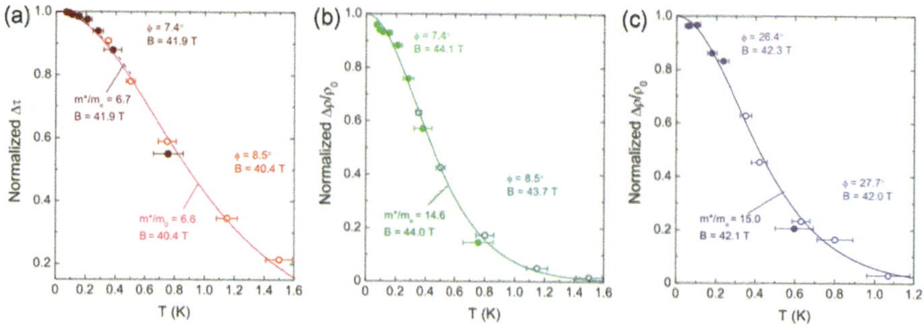

Fig. 3.4 Temperature dependence of the normalized amplitude of **a** the dHvA effect and **b**, **c** the SdH effect in the insulating phase of YbB_{12}. The solid curves in each panel and the dotted curve in (**a**) are LK fits. The corresponding effective mass is also given in each panel as a fitting parameter.

different bands. The frequency of dHvA oscillations has a 2D-like nature: F can be well fitted by the cylindrical Fermi surface model expressed as Eq. 2.16. However, the dHvA signal disappears when the field is applied at an angle above $\phi \sim 20°$. The relatively weak angle dependence within the small angle range of $|\phi| < 20°$ cannot rule out the possibility that the Fermi surface has a more prominent 3D nature. The frequency obtained using the SdH signal, on the other hand, shows a non-monotonic angle dependence. In the small angle range of $|\phi| < 20°$, the SdH frequency shows a stronger angle dependence than that expected from the 2D model. Furthermore, above $\sim 20°$, F decreases with increasing ϕ. The overall angular dependence can be explained by the 3D Fermi surface illustrated in Fig. 3.5b but not the 2D cylindrical model. The steep change in the small angle range can be traced by assuming a hyperboloid model, which is expressed as

$$F(\phi) \propto (\cos^2 \phi - r \sin^2 \phi)^{-1/2} \tag{3.1}$$

with the shape scaling parameter $r = a^2/c^2$ in the hyperboloid equation $(x^2 + y^2)/a^2 - z^2/c^2 = 1$. The black dashed curve in Fig. 3.5a shows the fitting result with the parameter $r = 7.42$. This hyperbolic Fermi surface is typically found in the "neck" regime connecting large Fermi surfaces, as shown in Fig. 3.5b. The frequency drop above $\phi \sim 20°$ can also be explained within this model by assuming large oblate 3D spheroids. The magenta dash-dot curve in Fig. 3.5a is a rough description of the high-angle data with the aspect parameter $b/a = 0.101$. The corresponding minimum and maximum cross-sectional areas perpendicular to the [100] crystalline axis for the neck and oblate spheroid are $A = 7.62$ and $37.5\ \mathrm{nm}^{-2}$, respectively. These 3D features of the Fermi surface revealed by the SdH effect strongly suggest that the SdH signal arises from the insulating bulk, rather than the (topological) 2D surface state.

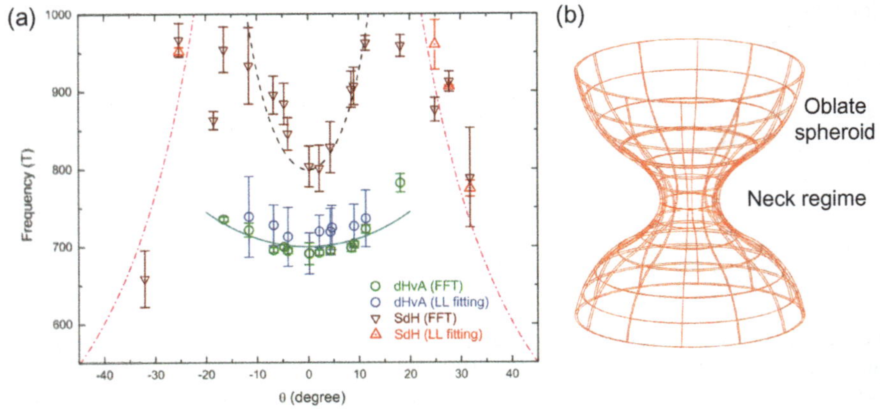

Fig. 3.5 **a** Angular mapping of the QO frequency obtained from the dHvA effect (open circles) and SdH effect (open triangles). The points were obtained using either FFT or Landau level fitting. The green solid curve is a fit obtained using a 2D Fermi surface model (Eq. 2.16) with $F_0 = 700\,T$. The black dashed curve represents the simulation of the neck-shaped Fermi surface model, while the magenta dash-dot curve shows the calculation with the oblate spheroid model. **b** Schematic of the 3D Fermi surface, which is consistent with the observations. © 2021 The author

3.4 High-Field Exotic Metal of YbB_{12}

3.4.1 SdH Effect in the Kondo Metal State

The observations of QOs in YbB_{12} imply the existence of a Fermi surface with charge-neutral fermions. To understand the nature of the QOs, it is informative to study how the electronic structure evolves and enters the field-induced Kondo metal (KM) phase as the energy gap closes. Moreover, the electronic properties realized in the field-induced metallic state are not fully understood, because the experimental techniques we can utilize are limited owing to the large critical field. In this section, we present transport measurements by using the proximity detector oscillator (PDO) technique in the KM phase to observe the SdH effect and discuss how the possible Fermi surface of the neutral fermions is affected by interactions with conventional charged fermions.

We employed the contactless PDO technique, which allows one to measure the conductivity of a metallic sample through the skin depth in a pulsed magnetic field environment [6, 7]. The shift in the resonant frequency f of the PDO circuit is given by

$$\Delta f = -a\Delta L - b\Delta R, \tag{3.2}$$

where L and R are the inductance and resistance of a sample coil, and a and b are constants. The change in L in the coil filled with the metallic sample is related to the change in the skin depth λ by $\Delta L \propto (r - \lambda)\Delta\lambda$, where r is the sample radius. The skin depth is also expressed by a combination of the resistivity ρ, angular frequency of the circuit ω, and permeability of the sample μ as

$$\lambda = \sqrt{\frac{2\rho}{\omega\mu}}. \tag{3.3}$$

Therefore, the frequency shift Δf is a measure of the MR of the sample. When the sample is insulating, the skin depth becomes as large as the sample radius, leading to the complete penetration of the field through the sample. In this case, the change in λ in the sample cannot be detected through Δf. This implies that the PDO technique is applicable only to metallic samples. Therefore, in the KI state of YbB_{12}, Δf is determined by the MR of the copper wire of the coil.

The field dependence of the PDO frequency f is shown in Fig. 3.6a for various field directions θ, which is the angle from [100] toward [110]. The I-M transition is manifested as the dip feature in f, as the sample skin depth shows a dramatic change across the critical field. On top of the positive MR on the KM phase, SdH oscillations are clearly detected, especially for small angles near $\theta{\sim}0°$. Figure 3.6b shows the oscillatory component of f, which was obtained by polynomial background subtraction, as a function of $1/H$. The steep change in the oscillation pattern above $\theta{\sim}20°$ is reminiscent of the previous observations of a dramatic change in the QOs in the KI phase at this angle (the dHvA effect disappears, and the SdH frequency starts to show a negative angular dependence). This similarity across the I-M transition implies that the QOs in both the insulating and metallic phase are determined by the same band. We discuss this point in more detail later.

The frequency pattern observed in the KM phase shows unusual behavior. Figure 3.6c shows the Landau level plots for the SdH oscillations in both the KM and KI states. The symbols in the KM phase are determined by the peaks and valleys, as labeled in Fig. 3.6b. Here, the + and − signs represent the spin-split Landau sublevels. Thus, the average of these characteristic fields corresponds to the oscillations from the spin-degenerate Fermi surface. The Landau index N in the KI phase shows the usual behavior of varying in proportion to $1/H$, and the slope corresponds to the field-independent quantum oscillatory frequency. On the other hand, it is remarkable that the Landau index in the KM phase does not show such behavior, implying that the SdH oscillations in KM are not periodic in $1/H$. This non-periodicity indeed leads to the failure of the FFT calculation for Δf: The summation of the calculated frequency components cannot reproduce the features in the raw data.

Nevertheless, despite the unusual periodicity, the temperature dependence of the SdH amplitude in the KM state can be well captured by the LK formulation (Eq. 2.13) as shown by the solid curves in Fig. 3.6d. The inset shows the field dependence of the cyclotron energy $E_c = eB/m^*$ obtained with each LK fit in the main panel. The H-dependence of E_c and the non-periodic SdH oscillations imply that the topology of the Fermi surface in the KM phase is H-dependent.

To analyze the non-periodic SdH oscillations in the KM state, we introduce an empirical relationship between the Landau index N and the characteristic magnetic

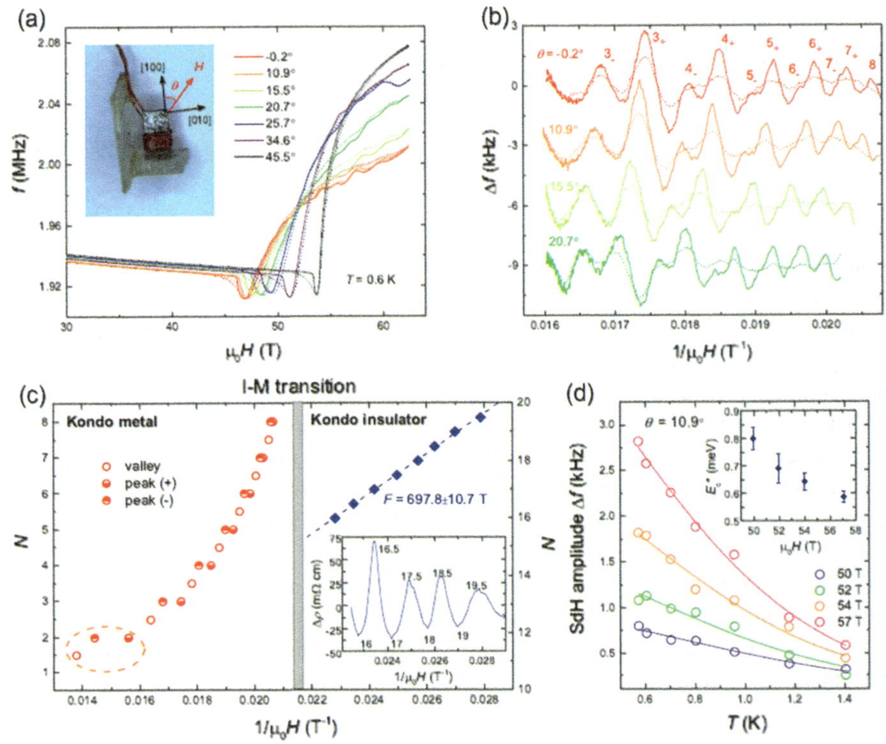

Fig. 3.6 **a** PDO frequency as a function of the field with various field directions. The inset shows an image of a sample wrapped by a PDO coil and defines the configuration of the angle θ. **b** Oscillatory component of the SdH signals in **a**. The integers are the Landau indices. The solid and dashed curves in **a** and **b** represent the data acquired from upsweeps and downsweeps, respectively. **c** Landau level plots for the KM and KI phases. The inset displays the SdH effect in the KI state. The data marked by the dashed orange circle are obtained using a 75 T Duplex magnet. **d** Temperature dependence of the SdH amplitude obtained at various magnetic fields. The inset shows the field dependence of the cyclotron energy determined from the fitting results in the main panel. © 2021 The author

field H_N at which the peaks or valleys are observed:

$$N + \lambda = \frac{F_0}{\mu_0(H_N - H^*)}, \tag{3.4}$$

where $\mu_0 H^*$ is an offset field, and λ is a phase factor. Setting $\mu_0 H^* = 41.6\,\mathrm{T}$, Eq. 3.4 successfully explains the relation between the characteristic H_N and Landau index N in the whole angle range below $\theta \sim 20°$, as shown in Fig. 3.7a. Owing to the offset term, Eq. 3.4 implies that the QO frequency is field-dependent. By setting $B \approx \mu_0 H$ and $B^* \approx \mu_0 H^*$, we introduce the B-dependent frequency as follows:

$$F_{\mathrm{KM}} = \frac{F_0}{B - B^*} B. \tag{3.5}$$

As discussed in Sect. 2.2, Onsager's rule still holds even if the cross-sectional area of the Fermi surface is B-dependent, i.e., $F(B) = A(B)\hbar/(2\pi e)$. Therefore, one can interpret the B-dependent frequency as a cross-sectional area that progressively depopulates as $A(B) = A_0 B/(B - B^*)B$, where $A_0 = 2\pi e F_0/\hbar$.

From the above perspective, the results indicate that the Fermi pockets in the KI and KM states, which are attributed to the QOs in those states, are identical or at least closely related. The field dependence of the frequency F_{KM} in the KM state for $H \parallel [100]$ is shown in Fig. 3.7b. A cursory inspection reveals that F_{KM} is quite close to the B-independent frequency in the KI state when F_{KM} is extrapolated back to $H_{\mathrm{I\text{-}M}}$. This coincidence of the frequency around the critical field $B = \mu_0 H_{\mathrm{KM}}(\theta)$ holds for all θ at which the SdH effect was observed. The magenta points in the inset of Fig. 3.7b shows the field dependence of the frequency, which is extrapolated from the KM state to the critical field. $F_{\mathrm{KM}}(\mu_0 H_{\mathrm{I\text{-}M}})$ closely traces that in the KI state, albeit with an offset of ~100 T. This offset might originate from the discontinuous jump of the cross-sectional area at the phase boundary or from the potential uncertainty of the critical field associated with a valence change [8]. Indeed, a slight offset to the critical field $\mu_0 H_{\mathrm{I\text{-}M}} - 0.8$ T is sufficient for the extrapolated frequency to trace well the angular dependence in the KI state (the red points in the inset of Fig.3.7b).

QOs also provide information on the cyclotron mass m^* of the quasiparticles. The comparison of m^* between the KI and KM states gives further supporting evidence that the Fermi surface attributed to the QOs is closely related in both phases. Figure 3.7c shows the field dependence of the cyclotron mass ratio in both the KI and KM states. Around the critical field, the m^* values for both phases coincide at ~7, which implies that the identical Fermi pocket is the source of the QOs in both states. As discussed previously, m^* increases with increasing H in the KM state. This mass enhancement is also captured by the spin-splitting parameter S induced by the Zeeman splitting of the Fermi surface. According to [9], the split spin-down (-up) Landau level reaches the Fermi level under the following condition:

$$\frac{F}{B_N^{\pm}} = N + \lambda \pm \frac{1}{2}S. \tag{3.6}$$

Here, S is determined from the g-factor and cyclotron mass as

$$S = \frac{1}{2}\frac{m^*}{m}g. \tag{3.7}$$

Thus, by tracing the additional phase factor in QOs owing to the Zeeman splitting, one can estimate S from the nonlinear (Fig. 3.6c) and linear (Fig. 3.7a) Landau level plots as $S = F_m(1/B_N^+ - 1/B_N^-)$ and $S = F_0(1/(B_N^+ - B^*) - 1/(B_N^- - B^*))$, respectively. Figure 3.7d depicts the field dependence of S calculated using both the formulas. S and m^* are related to Eq. 3.7 and follow the same field dependence on substituting $g = 0.084$ (the red points in Fig. 3.7d). The coincidence of the field dependence not only validates the Zeeman splitting analysis, but also reveals the exotic property of the KM state, which hosts an unusually small g-factor compared to a non-interacting electron system ($g = 2$).

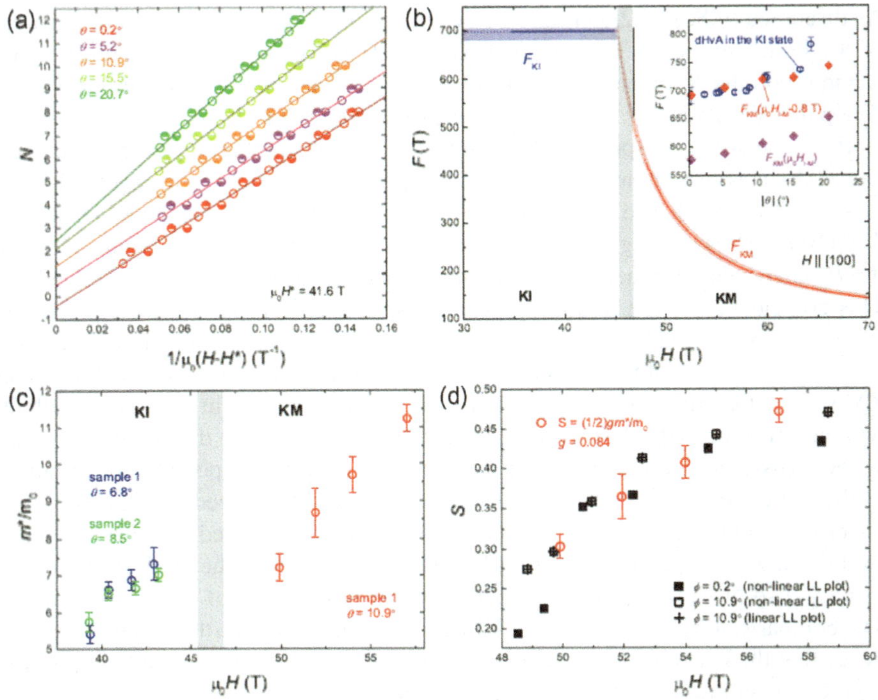

Fig. 3.7 **a** Landau level plots as a function of $1/\mu_0(H - H^*)$. The offset field $\mu_0 H^* = 41.6\,\mathrm{T}$ is determined so that the plot becomes linear. The slope for each line represents F_0 in Eq. 3.5. **b** Field dependence of the QO frequency F. F is constant in the KI state but becomes B-dependent in the KM state, as described by Eq. 3.5. The solid and dashed lines represent the field range where the SdH oscillations are present and absent, respectively. The colored regime along each curve represents the width of the FFT peak. The black line around the phase boundary denotes the maximum allowed mismatch between F_{KI} and F_{KM} of $\sim$180 T. The inset shows the field dependence of F. **c** Field dependence of the cyclotron mass ratio m^*/m. **d** Field dependence of the spin-splitting parameter S. The red points are estimated from the cyclotron mass in the KM state with $g = 0.084$.

3.4.2 Fermi Liquid Behavior and Two-Fluid Picture

The exotic properties of KM are also revealed by magnetotransport experiments. To measure the resistivity in the KM state reliably, the pulsed-current technique was employed. As the excitation current produces Joule heating in the KI state, in which the resistivity is relatively large, the current was applied only in the KM state. The time profiles of the current and magnetic fields are shown in the inset of Fig. 3.8a. A current pulse was applied so that one can measure transport only when $H > H_{\mathrm{I-M}}$. Figures 3.8a, b show the temperature dependence of the observed resistivity $\rho(T)$ at 55 T, which is obtained by performing the pulsed-current measurement at various temperatures. Although the hybridization gap is completely suppressed in the KM state, the $\rho(T)$ curve shows typical behavior for Kondo lattice systems: $\rho(T)$ shows a peak at a coherent temperature $T^* = 14$ K and decreases as T decreases, reflecting the formation of coherent heavy quasiparticle band. A remarkable feature found in the KM state is that it shows a temperature-linear resistivity, i.e., $\rho \propto T$, below T^*, which is the hallmark of non-FL behavior [10]. This anomalous behavior, however, only holds within the temperature range of $T_{FL} < T < 9$ K, where $T_{FL} = 2.2$ K is the temperature below which FL behavior is observed ($\rho \propto T^2$). Figure 3.8b shows the T^2 dependence of ρ below $T^2 = 10$ K. There is a good agreement between the data and Eq. 1.10, and $\rho_0 = 0.34$ mΩcm and $A_2 = 58$ $\mu\Omega$cm were obtained by fitting, as indicated by the dotted line in Fig. 3.8b. As the residual resistivity ρ_0, in general, originates from impurity scattering, this relatively large ρ_0 indicates that the KM state may be classified as a "bad metal."

The Fermi surface attributed to the QOs does not contribute to the charge transport in KM for the following reason. The observations of the B-dependent QO frequency and cyclotron mass imply that the geometry of the Fermi surface in the KM state significantly changes and most likely shrinks by $\simeq$45% from 50 T to 60 T, according to Eq. 3.5 and B-dependent Onsager's rule. Assuming a spherical Fermi surface, this shrinkage corresponds to a $\simeq$60% reduction in the quasiparticle density n, whereas the observed cyclotron mass increases by $\simeq$60%. Consequently, the Drude expression $\rho = m^*/ne\tau$ predicts that the resistivity would increase by a factor of 4 from 50 T to 60 T. In contrast, the observed MR in the KM state is negligibly small, indicating that the Fermi surface revealed by the SdH effect is charge-neutral even in the KM state. To account for the charge-transport properties in the KM state, one might have to assume another Fermi pocket with conventional charged fermions. This hidden Fermi pocket is also necessary to explain the large Sommerfeld coefficient observed in the pulsed-field specific heat measurement [11], which yields $\gamma \sim 63$ mJ/molK^2 on interpolating the value at 55 T. First, we estimate the contribution to γ from the Fermi surface detected by the SdH effect in the KM state as follows. Assuming the simplest case of a single 3D band of the isotropic FL, the Sommerfeld coefficient is given by

$$\gamma = \frac{\pi^2 k_B^2}{3} \frac{m^* k_F}{\pi^2 \hbar^2}, \tag{3.8}$$

where k_F is the Fermi vector. Inserting m^* and k_F, which are estimated from the SdH analysis, we find that γ from this Fermi pocket makes a contribution of only 4.4% of the measured γ. This rough estimation indicates that the γ is determined not only by the Fermi pocket resolved from the SdH effect, but also by the other Fermi surface of conventional charged fermions that we miss in the QOs.

The implication of the hidden Fermi surface in the KM state can be also captured by the unusually large Kadowaki–Woods (KW) ratio. By combining $\gamma \sim 63$ mJ/molK^2 from the pulsed-field specific heat measurement [11] and $A_2 = 58$ $\mu\Omega$cm from the previous linear fitting of $\rho(T)$ below T_{FL} (Fig. 3.8b), we estimate the KW ratio in the KM state of YbB_{12} at 55 T as $1.46 \times 10^{-2} \mu\Omega$cm(Kmol/mJ)^2, which is 3–4 orders of magnitude larger than the universal value given in Eq. 1.11. In Fig. 3.8c, we represent the relation between A_2 and γ^2 in the KM state of YbB_{12} along with those of other classes of materials including transition metals, Ce- and U-based heavy fermions,

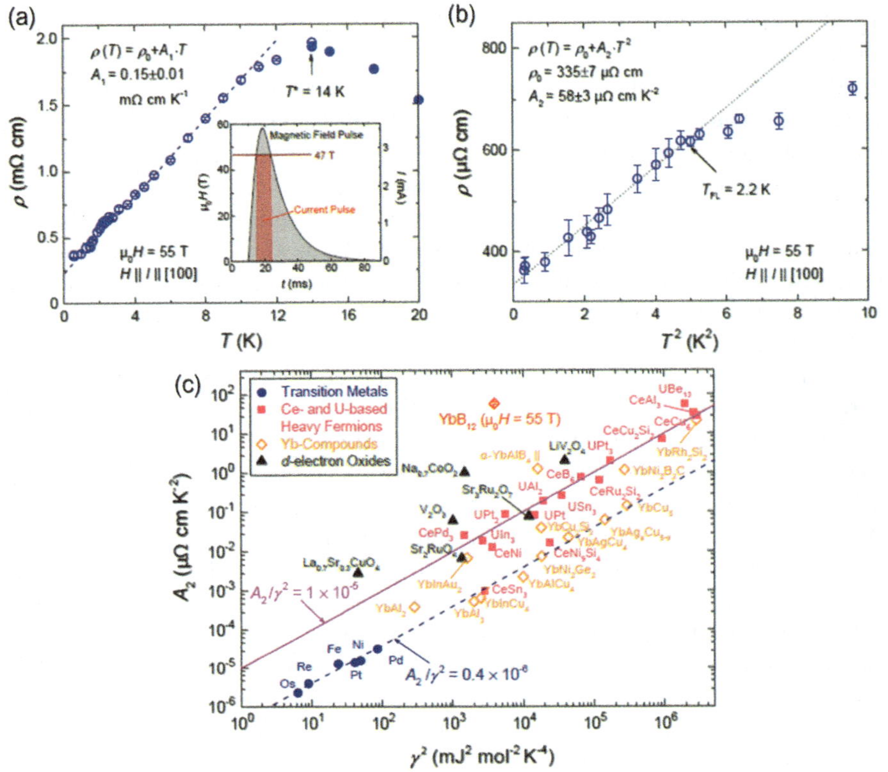

Fig. 3.8 **a** Temperature dependence of resistivity of the KM state of YbB_{12} at 55 T. The dashed line is the fitting for the T-linear resistivity regime. The inset shows the time profile of the pulsed magnetic field and current pulse. **b** Resistivity plotted against T^2. The dashed line is the fitting below T_{FL}, and the parameters are also shown. **c** Kadowaki–Woods plot for a wide variety of materials, including transition metals (indigo circles), Ce- and U-based heavy fermions (magenta squares), Yb-based compounds (orange diamonds), and d-electron oxides (black triangles). The KM state of YbB_{12} is plotted as a red diamond. © 2021 The author.

Yb-compounds, and d-electron oxides. This strong deviation in the KM state verifies the exotic properties of KM, and one can estimate information on the hidden Fermi surface (k_F and m^*) by combining the unusually large A_2 and γ. Employing the same calculations as in [12], A_2 can be written as

$$A_2 = \frac{81\pi^3 k_B^2}{4e^2\hbar^3} \frac{m^{*2}}{k_F^5}. \tag{3.9}$$

Combining Eqs. 3.8 and 3.9, one can estimate $k_F = 2.15$ nm^{-1} and $m^* = 90.0 m_0$. Although such a large effective mass is unusual in Yb-based compounds, it can explain why the charged fermions cannot contribute to the QOs in the current environment of the pulsed-field measurements.

The results imply the coexistence of two fluids in KM state: (i) charge-neutral fermions (CNFs) and (ii) the more conventional but still exotic charged fermions. The CNFs contribute to the QOs in both the KI and KM states, but are not responsible for charge transport. The existence of the CNFs even in the KM state is supported by the similarity of the Fermi surface at $H_{\text{I-M}}$ and little contribution to MR, although the geometry of the Fermi pocket shows significant shrinkage and mass enhancement with increasing field. The emergence of the charged fermions above $H_{\text{I-M}}$ is supported by the unusually large γ observed in the specific heat measurement and the FL behavior with the large coefficient A_2, which strongly violates the KW ratio. Based on the two-fluid picture, the I-M transition produces a sudden increase of the density in (ii), which becomes a dense liquid of heavy fermions in the KM state from a thermally excited low-density gas in the KI state. On further increasing B in the KM state, (i) becomes less energetically favorable, and the Fermi surface of CNFs shrinks, as described by Eq. 3.5. The hidden Fermi surface of (ii) acts as a "reservoir" into which CNFs can scatter or transfer. The analogous situation of the two-fluid picture has also been reported in other materials [13, 14].

The exotic metallic state found in the high-field regime involves the survival of the CNFs above the gap closure and their coexistence with the charged FL. This peculiar two-fluid picture may explain the observed non-FL behavior. In this situation, Luttinger's theorem may be violated, leading to possible continuous variation of FL properties [15]. In addition, the T-linear resistivity may be attributed to the interaction between CNFs and charged fermions [16]. The exotic properties of the KM state we found, therefore, are a prerequisite for future theoretical works.

3.5 Conclusion

We reviewed the high-field electronic properties of high-quality single crystals of the KI YbB_{12} and successfully observed QOs both in the KI state and KM state. Remarkably, YbB_{12} exhibits both the dHvA effect and SdH effect in the insulating phase well below the I-M transition field, suggesting that the QOs are an intrinsic property of the insulating ground state. Indeed, the amplitude of the SdH signals increases as the activation gap increases, ruling out the possibility that a metallic impurity phase contributes to the QOs. Furthermore, the temperature damping factor shows good agreement with the LK formula for a relatively large effective mass $m^* \sim 6\text{-}15 m_0$, demonstrating that the quasiparticles contributing to the QOs follow the Fermi–Dirac distribution law. The large observed m^* also indicates that the electron correlation plays an important role in realizing the quasiparticles. Although the angular mapping of the frequency obtained from the dHvA effect cannot rule out the possibility of a 2D SS, the angle-resolved SdH signals strongly suggest that the Fermi surface has a 3D-like nature. The observations provide intriguing evidence that the KI YbB_{12} hosts a Fermi surface, which is a defining characteristic of metals. This apparent contradiction leads to the exotic Fermi surface of neutral fermions, which is also supported by the thermal transport experiment that we discuss in Chap. 5. The observation of QOs in the insulating phase of YbB_{12} makes a distant allusion to the fact that the QOs may be a common feature among KIs, but further investigations on SmB_6 and exploration on other types of KIs are required to confirm the universality.

We also examined the SdH signals above the critical field by employing the PDO technique and captured the exotic properties of the field-induced metallic state. Although the SdH effect in the KM state shows a nonlinear Landau level plot, our analysis reveals a similarity between the Fermi surfaces in the KI state and KM state. We concluded that the Fermi pocket contributing to the QOs in the KM state is identical to that in the KI state. The results can be, therefore, interpreted in terms of a two-fluid picture: the coexistence of (i) CNFs, which contribute to the QOs in both the KI and KM states, and (ii) the more conventional but still exotic heavy FL. Although the CNFs are not strongly affected at the I-M transition, they become energetically unfavorable and convert into (ii) as the field increases. The field evolution of the Fermi surface of (i) is indeed captured by the unusual field suppression of F_{KM} and enhancement of m^*. Although the Fermi surface of (i) can be resolved by the current environment of pulsed-field measurements, it is not the case for (ii) owing to the surprisingly large $m^* \approx 90m_0$. Despite the absence of QOs from (ii), this hidden Fermi surface contributes to the large γ and extremely massive A_2, which strongly violates the universal Kadowaki–Woods ratio. In addition, analysis on the spin-splitting QOs enabled us to evaluate the extremely small g ~0.084 of the neutral fermions. The coexistence of the two fluids and their interaction may explain the non-FL behavior found in the KM state. The results provide strong constraints for future theoretical investigations to explain the origin of CNFs and exotic metallic properties.

References

1. Xiang Z, Kasahara Y, Asaba T, Lawson B, Tinsman C, Chen L, Sugimoto K, Kawaguchi S, Sato Y, Li G, Yao S, Chen YL, Iga F, Singleton J, Matsuda Y, Li L (2018) Science 69:65
2. Xiang Z, Chen L, Chen KW, Tinsman C, Sato Y, Asaba T, Lu H, Kasahara Y, Jaime M, Balakirev F, Iga F, Matsuda Y, Singleton J, Li L (2021) Nat Phys 17:788
3. Iga F, Shimizu N, Takabatake T (1998) J Magn Magn Mater 177–181:337
4. Iga F, Hiura S, Klijn J, Shimizu N, Takabatake T, Ito M, Matsumoto Y, Masaki F, Suzuki T, Fujita T (1999) Physica B 259–261:312
5. Liu H, Hartstein M, Wallace GJ, Davies AJ, Hatnean MC, Johannes MD, Shitsevalova N, Balakrishnan G, Sebastian SE (2018) J Phys Condens Matter 30:16LT01
6. Altarawneh MM, Mielke CH, Brooks JS (2009) Rev Sci Instrum 80:066104
7. Ghannadzadeh S, Coak M, Franke I, Goddard PA, Singleton J, Manson JL (2011) Rev Sci Instrum 82:113902
8. Götze K, Pearce MJ, Goddard PA, Jaime M, Maple MB, Sasmal K, Yanagisawa T, McCollam A, Khouri T, Ho P-C, Singleton J (2020) Phys Rev B 101:075102
9. Shoenberg D (1984) Cambridge University Press. England, Cambridge
10. Varma CM (2019) Rev Mod Phys 92:31001
11. Terashima TT, Matsuda YH, Kohama Y, Ikeda A, Kondo A, Kindo K, Iga F (2018) Phys Rev Lett 120:257206
12. Jacko AC, Fjærestad JO, Powell BJ (2009) Nat Phys 5:422
13. Matsumoto Y, Kuga K, Tomita T, Karaki Y, Nakatsuji S (2011) Phys Rev B 84:125126

14. Harrison N, Singleton J, Bangura A, Ardavan A, Goddard PA, McDonald RD, Montgomery LK (2004) Phys Rev B 69:165103
15. Senthil T, Vojta M, Sachdev S (2004) Phys. Rev. B 69:035111
16. Patel AA, McGreevy J, Arovas DP, Sachdev S (2018) Phys Rev X 8:021049

Chapter 4
Topological Surface Conduction in YbB_{12}

4.1 Introduction

KIs are a class of strongly correlated electron systems, which have a long research history over the past fifty years. There, hybridization between localized f-electron and conduction electron band gives rise to opening of an insulating gap at the Fermi level at low temperatures [1, 2]. Clear signatures of a Kondo hybridization gap have been reported in KIs by various measurements. KIs have received renewed interest in recent years [3]. It has been suggested theoretically that some KIs are topological insulators which possess topologically protected metallic two-dimensional (2D) SS. Until now, topological insulator states have been mostly studied in non-correlated band insulators, in which the band inversion is induced by the spin-orbit interactions. In KIs, on the other hand, the band inversion develops through the interplay of strong electron correlations and spin-orbit interactions [4]. Thus electron correlation effects are essentially important for the formation of the topologically non-trivial state. The TKI is a fascinating realization of the 3D topological insulator state in strongly correlated electron systems.

It has been suggested theoretically that SmB_6 is a strong topological insulator, which exhibits an odd number of surface Dirac modes characterized by a Z_2 topological index protected by the time-reversal symmetry [5]. Recently, the resistivity plateau in SmB_6 has been attributed to the topologically protected 2D SS. Indeed, a series of transport measurements in SmB_6 provides evidence for the presence of SSs [6–8]. A salient feature of the topological surface metallic state is a helical spin texture due to spin-momentum locking. In SmB_6, such a spin-momentum locking has been directly observed by spin-polarized angle-resolved photoemission spectroscopy (ARPES) [9]. According to the band calculations, YbB_{12} is a topological crystalline KI characterized by a non-zero mirror Chern number protected by the crystalline reflection symmetry [10]. In YbB_{12}, however, the nature of the surface metallic state has been little explored, although a SSs has been observed by ARPES [11].

Recently, other salient aspects of SmB_6 and YbB_{12} have been reported. In an external magnetic field, both compounds exhibit QOs despite the opening of significant charge gaps [12–15]. As the QOs are a signature of the Landau quantization of the

Y. Sato, *Quantum Oscillations and Charge-Neutral Fermions in Topological Kondo Insulator YbB_{12}*, Springer Theses,
https://doi.org/10.1007/978-981-16-5677-4_4

Fermi surface, the observations have come as a great surprise. In particular, in YbB_{12}, QOs are observed not only in the magnetization (de Haas–van Alphen effect) but also the resistivity (Shubnikov–de Haas, SdH, effect). The 3D character of the SdH signals in YbB_{12} suggests that the QOs arise from the electrically insulating bulk [14, 16]. To explore the peculiar properties of YbB_{12}, we need more detailed information about the electronic structure, particularly the nature of the surface conduction.

In this chapter, we present the results of transport measurements on YbB_{12} single crystals and microstructures with controlled thickness [17]. The magnitude of resistivity plateau at low temperatures decreases linearly with the sample thickness, indicating the presence of a surface conduction. Moreover, in the micostructued device where the surface conduction is dominant, we observe the weak-antilocalization (WAL) effect in small magnetic fields, as reported in other topological insulators. These results suggest a signature of spin-momentum locking in the SS due to topological protection. Our results thus provide a supporting evidence of the topologically protected SS in YbB_{12}.

4.2 Experimental

High-quality YbB_{12} single crystals were grown by the traveling-solvent floating zone method [18]. The crystals were cut from as-grown ingots and polished into a rectangular shape (A1, 2, and 3). The microstructures (B1, 2, and 3) were fabricated by focused ion beam (FIB) method to control the thickness of the single crystals in micron size. The fabrication process for the microstructures is described in Fig. 4.1a. First, we cut a cuboid lamella from A2 single crystal by Ga^{2+} ion beam. The edges of the lamella are oriented along [100] direction and its equivalent crystalline axes. This lamella was then transferred to a Si/SiO_2 substrate and glued using an epoxy. To make electrical contacts, the lamella was sputter-coated by 100 nm gold, which was partly removed later by FIB etching from the top surface. The images of the prepared microstructural devices taken by scanning electron microscopy are represented in Fig. 4.1b, c. For B2 and B3 microstructures, the lamella was further patterned into the thin bar shape by using FIB.

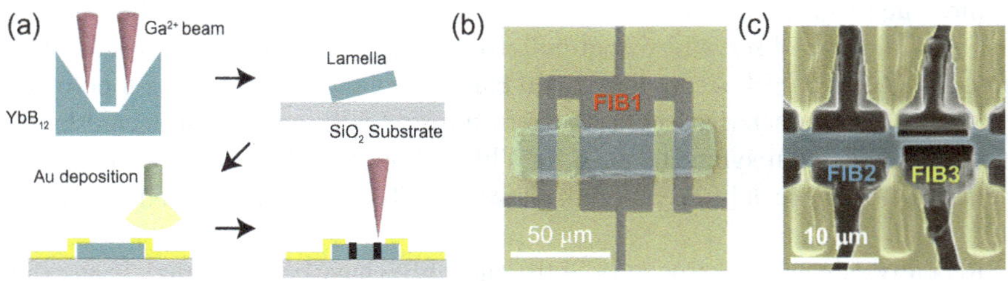

Fig. 4.1 **a** Schematic of the fabrication process for the microstructure devices. **b, c** Scanning electron microscope images of microstructures fabricated by FIB. YbB_{12} crystal, gold contacts, and Si/SiO_2 substrate are colored in blue, yellow, and gray, respectively. © 2021 The author

Table 4.1 Dimensions [length (ℓ), width (w), and thickness (t)] and fitting parameters obtained by using Eq. 4.1 of the single crystals (A1-3) and microstructures (B1-3)

Sample	Dimension (mm)			ρ_b^1 ($\mu\Omega$cm)	ρ_b^2 (mΩcm)	Δ_1 (meV)	Δ_2 (meV)	ρ_n (ΩcmK)
	ℓ	w	t					
A1	6.84	0.52	0.51	74.0	1.30	11.2	4.23	0.0274
A2	1.21	0.95	0.20	74.5	1.25	11.1	4.14	0.0436
A3	1.78	1.82	0.13	78.1	1.72	11.0	3.50	0.117
B1	37.5×10^{-3}	19.1×10^{-3}	8.2×10^{-3}	91.0	1.71	10.9	4.15	1.96
B2	8.45×10^{-3}	2.6×10^{-3}	1.0×10^{-3}	95.8	2.90	11.2	3.50	1.96
B3	8.45×10^{-3}	1.0×10^{-3}	0.6×10^{-3}	92.6	3.01	11.1	3.33	2.01

Dimensions of all the samples used in this study are listed in Table 4.1. The sample thickness t is defined as a short side in the cross section rectangle. The electrical contacts were made using silver paste for the single crystals and by evaporated gold for the microstructures. The resistivity measurements were performed using a standard four-contact configuration. High-field resistivity measurements were performed using a capacitor-driven 65 T pulsed magnet at the National High-Magnetic Field Laboratory, Los Alamos, and the data were taken via a high-frequency a.c. technique with a specialized digital lock-in program. The driving signal ($f = 75$ kHz) was generated by an a.c. voltage source and applied to the sample across a transformer. The current through the sample was monitored using a 1 kΩ shunt resistor and was determined to be 12.4 μA.

4.3 Metallic Surface Conduction in YbB_{12}

Fig. 4.2 depicts the temperature (T) dependence of the resistivity ρ of the YbB_{12} single crystals and microstructures plotted on a log–log scale. In all crystals and microstructures, $\rho(T)$ increases by several orders of magnitude from room temperature with decreasing T, which is attributed to the bulk insulating channel. On further lowering the temperature, $\rho(T)$ becomes weakly T-dependent and shows a resistivity plateau, as observed in SmB_6 [6–8]. Although $\rho(T)$ above 30 K well coincides for all samples, the magnitude of the resistivity plateau is strongly suppressed with reducing t. These results indicate that the resistivity plateau is attributed to the surface metallic states.

Figure 4.3a shows the Arrhenius plot of $\rho(T)$ for A1 single crystal. The slope of the plot changes at around $T \approx 14$ K ($1/T \approx 0.7$ K^{-1}), showing a two-gap behavior. Fitting with a thermal activation model of resistivity ($\rho(T) \propto \exp(\Delta/2k_B T)$), we

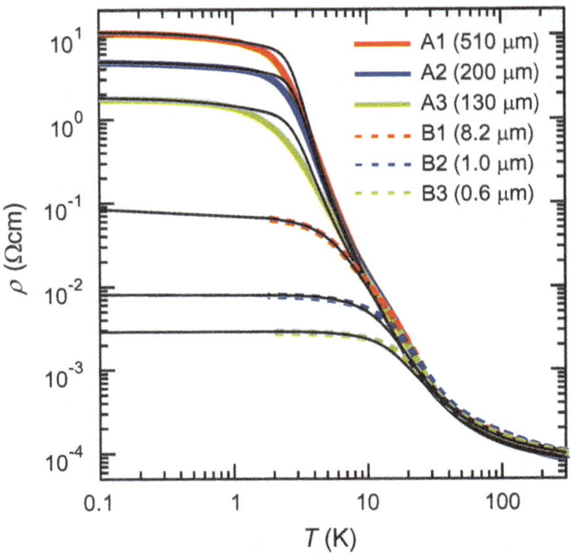

Fig. 4.2 Temperature dependence of resistivity ρ. The solid and dotted lines are the results of single crystals (A1-3) and microstructures (B1-3), respectively. The black curves are the fits by Eq. 4.1. © 2021 The author

obtain the gap magnitude of $\Delta_1 = 11.2\,\text{meV}$ and $\Delta_2 = 4.23\,\text{meV}$ within the temperature ranges $16\,\text{K} < T < 50\,\text{K}$ and $4.5\,\text{K} < T < 8.5\,\text{K}$, respectively, in agreement with the previously reported values [14, 18].

A parallel conduction model is used to extract the bulk and surface contributions. The total resistivity is given by

$$\frac{1}{\rho} = \sum_{i=1,2} \frac{1}{\rho_{\text{b}}^{i}} \exp\left(-\frac{\Delta_i}{2k_{\text{B}}T}\right) + \frac{1}{\rho_{\text{s}}} + \frac{T}{\rho_{\text{n}}}. \tag{4.1}$$

The first term in the right hand side is the contribution arising from the insulating bulk channel with two characteristic gaps Δ_i $(i = 1, 2)$ as discussed above. The second term represents the contribution arising from the surface metallic state. In this model, we first pay attention to its t-dependence, and we assume that this contribution is temperature-independent [19, 20]. The third term is a T-linear term, which has been reported in SmB_6 and discussed in terms of a possible nodal semimetallic state originating from defects in a crystalline lattice [21, 22]. It has been reported that in SmB_6, ρ_{n} is independent of the sample thickness, indicating the bulk origin [21]. In YbB_{12}, on the other hand, ρ_{n} largely depends on the thickness as shown in Table 4.1. Although the origin of ρ_{n} is not well understood, $\rho(T)$ for all samples can be well fitted by Eq. 4.1, as shown by the solid black lines in Fig. 4.2. It should be noted that for the bulk channels, the activation gaps Δ_i as well as the bulk resistivity ρ_{b}^{i} are nearly unchanged by reducing t. This demonstrates that the bulk electronic conduction is little damaged during the FIB process, in which a convergent gallium ion beam is irradiated onto a sample and a specific part is made thinner. In Fig. 4.3b, the magnitude of the surface resistivity ρ_{s} is plotted as a function of the sample thickness on a log–log

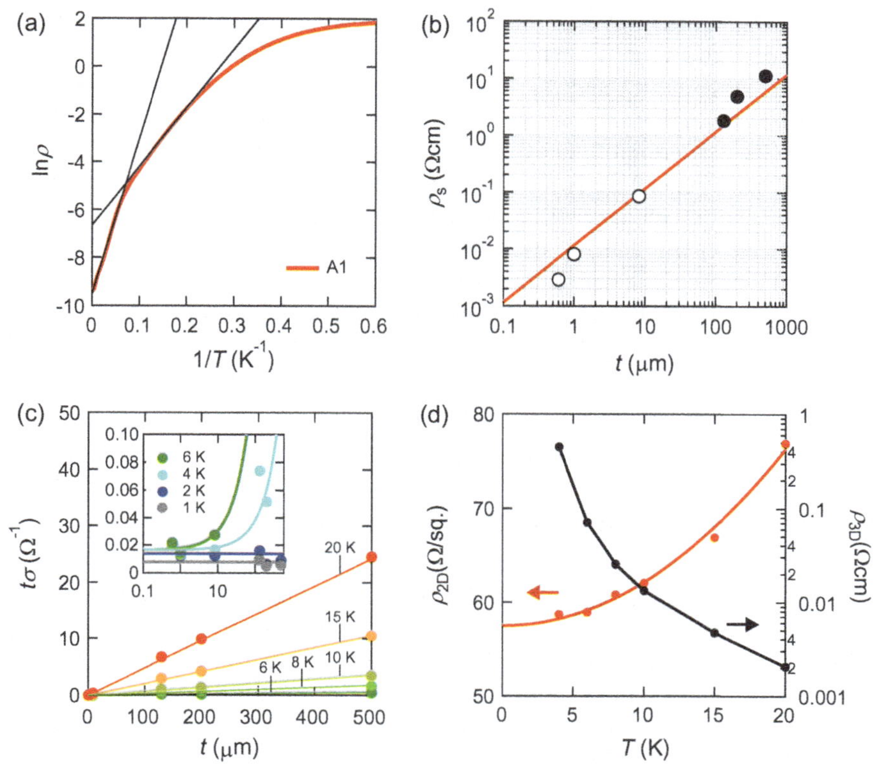

Fig. 4.3 **a** Arrhenius plot of the resistivity ρ for A1 single crystal plotted as a function of $1/T$. The solid black lines are the fits by the thermal activation model. **b** Thickness (t) dependence of the surface resistivity ρ_s obtained from the fitting of $\rho(T)$ to Eq. 4.1. The filled and open circles represent the results of the single crystals and microstructures, respectively. The red line is a linear ($\rho_s \propto t$) fit. **c** t dependence of conductance $t\sigma$ with different T. The lines are the linear fits. The inset shows the same plot below 6 K in log(t) scale. **d** Extracted T dependence of ρ_{3D} and ρ_{2D} obtained by the linear fit in (**c**). The red line shows T^2 fit, whereas the black one is a guild to the eyes. © 2021 The author

scale. As shown by the red line, the surface resistivity increases linearly proportional to t. This indicates that the low-temperature plateau values of the sheet resistance is independent of the sample thickness. Thus, the present results demonstrate that the SS is an intrinsic property and is robust against the irradiation in YbB_{12}. We note that in SmB_6 ion irradiation does not destroy the SS but produces a damaged layer that could be poorly conducting [23], resulting in a deviation from the linear relationship between ρ_s and t.

The presence of the metallic surface channel is further supported by its extracted T-dependence. Sample conductance $t\sigma$ can be expressed by using 2D (3D) resistivity ρ_{2D} (ρ_{3D}) as,

$$t\sigma = \frac{1}{\rho_{2D}} + \frac{t}{\rho_{3D}}. \tag{4.2}$$

Figure 4.3c shows the t dependence of $t\sigma$ with different temperatures obtained from the single crystals (A1-A3) and the microstructures (B1-3). By performing the linear fit, one can separate ρ_{3D} and ρ_{2D} at a given temperature. We note that we cannot proceed the analysis below $T \sim 4$ K, because the slope becomes negligibly small with respect to the scattering of the experimental data. Estimation of ρ_{2D} also becomes difficult when temperature is too high because the change in the second term in Eq. 4.2 becomes much larger than the first term. Figure 4.3d displays the extracted T-dependence of ρ_{3D} and ρ_{2D} over the temperature range $4\,K < T < 20\,K$, which clearly demonstrates the typical transport feature expected in topological insulators: coexistence of 3D insulating bulk and metallic surface. As shown by the red line in Fig. 4.3d, it is remarkable that ρ_{2D} shows Fermi liquid behavior, i.e., $\rho_{2D} \propto T^2$ with a small residual resistivity ~ 58 Ω/sq., which is comparable to that in graphene films grown by chemical vapor deposition [24].

4.4 Magneto Transport of the Microstructure

A signature of the spin-momentum locking provides further support for the topological SS. In the topological SS, the spin-momentum locking often gives rise to the WAL effect, which is caused by the destructive interference between time-reversed electron paths and lowers the sample resistance. This effect is suppressed by applying magnetic field that breaks time-reversal symmetry, which causes a dip in the magnetoresistance (MR) at zero field [25–27]. Such a WAL effect has been observed in SmB_6 [28] and other topological insulators such as Bi_2Te_2Se [29], which is attributed to the topological SSs. Figure 4.4 shows the 2D magnetoconductance $\Delta\sigma = \sigma(H) - \sigma(0)$ in low fields at 1.8 K for the B1 microstructure in which the surface contribution to the electrical transport is larger than the bulk. Magnetic field is applied perpendicular to the major plane of the crystal ($H \parallel t$ in Fig. 4.1a) so that the WAL effect is expected to be most pronounced. With increasing the magnetic field, $\Delta\sigma$ steeply decreases, which is consistent with the WAL effect. Above 1.5 T, $\Delta\sigma$ turns to increase with H. The positive magnetoconductance at higher fields is a typical feature of KIs, which is caused by the suppression of the hybridization gap by magnetic field [14, 30, 31].

In 2D electronic systems in the limit of long inelastic scattering time, WAL gives a negative quantum correction to the conductivity, which is described by Hikami–Larkin–Nagaoka (HLN) equation [32]:

$$\Delta\sigma = -\frac{\alpha e^2}{\pi h}\left[\psi\left(\frac{B_0}{B}+\frac{1}{2}\right) - \ln\left(\frac{B_0}{B}\right)\right] + \beta B^2, \tag{4.3}$$

where $\psi(x)$ is the digamma function, $B = \mu_0 H$ is perpendicular magnetic field component to the plane, and $B_0 = \hbar/(4eL_\phi^2)$ with the dephasing length L_ϕ. The B^2 term includes reduced spin-orbit and elastic scattering corrections as well as the classical cyclotronic contribution [29]. Taking into account the contributions from

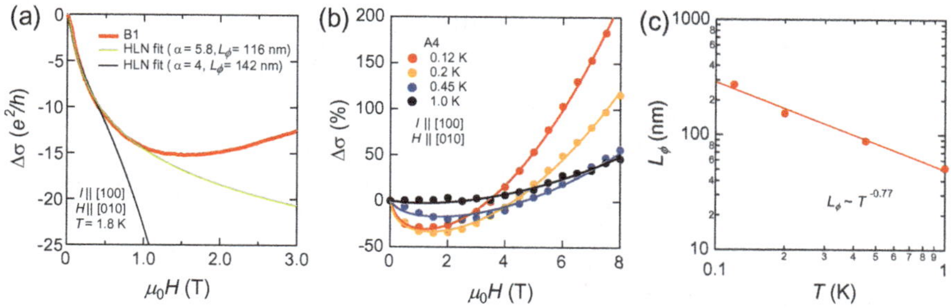

Fig. 4.4 **a** Magnetoconductance $\Delta\sigma = \sigma(H) - \sigma(0)$ at 1.8 K for the B1 microstructure. The solid thin yellow and black lines show fits to the Hikami–Larkin–Nagaoka (HLN) model (Eq. 4.3). In the latter case, α is fixed to 4. **b** Magnetoconductance of the A4 single crystal for different temperatures. The solid lines are the fits to the HLN model. **c** Temperature dependence of the dephasing length L_ϕ in the A4 single crystal. The line is the fit to $T^{-\gamma}$. © 2021 The author

both top and bottom surface side, $\alpha = 1$ is expected in a topological insulator with single Dirac band channel. In the presence of multiple Dirac bands contributing to the surface metallic conduction, the α value is larger than 1. As four Dirac points appear on the [100] surfaces in YbB_{12} [10], the total number of channels is eight, resulting in $\alpha = 4$ if four Dirac bands individually contribute to the surface conduction with similar L_ϕ values. The black line in Fig. 4.4a shows the fitting with $\alpha = 4$ and $L_\phi = 142$ nm. The low field data are well reproduced. The best fit is yielded by using $L_\phi = 116$ nm and $\alpha = 5.8$, as shown by the yellow line in Fig. 4.4a. The deviation at high field is due to the negative MR in KIs. We note that in some of 3D topological semimetals and insulators, α obtained from the WAL effect largely deviates from the expected values. This has been attributed to a contribution from bulk conducting channels and coherent surface-to-bulk scattering [28, 29, 33, 34]. Thus, the present result of α in the YbB_{12} microstructure, which is close to the expected value in the 2D system, indicates that the WAL effect arises from the SS in the presence of spin-momentum locking, providing supporting evidence of topological SSs in YbB_{12}.

The WAL effect is reproduced even in single crystals as long as they are in the surface conduction regime at low temperatures. Field dependence of magnetoconductance of another single crystal A4 is presented in Fig. 4.4b. The significant drop of the magnetoconductance at low fields is pronounced at low temperatures but it becomes invisible at 1 K. All the data are well fitted by the LHN model up to 8 T as shown by the solid lines in Fig. 4.4b. Temperature dependence of the dephasing length L_ϕ extracted by the fits is shown in Fig. 4.4c. L_ϕ decays as $T^{-\gamma}$ with $\gamma = 0.77$, which is comparable to γ reported in SmB_6 ($\gamma = 0.55$) [28] and Bi_2Te_2Se ($\gamma = 0.75$) [29]. This result suggests that a similar scattering mechanism plays a role in the topologically protected SSs.

To obtain the information on the SdH oscillations reported in bulk YbB_{12} [14], we measured the MR of the microstructure at high magnetic fields. Figure 4.5a shows the MR of A2 single crystal and B1 microstructure above 30 T at 0.5 K. In this field range, the MR of both single crystal and microstructure steeply decreases with

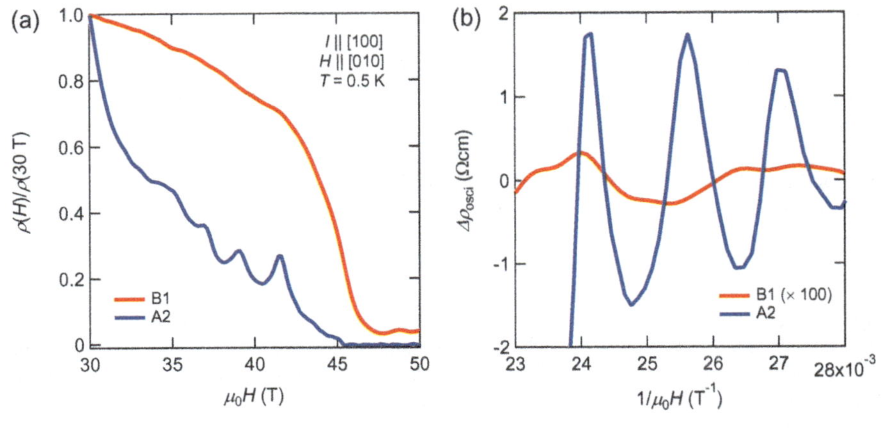

Fig. 4.5 **a** Magnetoresistance normalized by the resistivity at 30 T, $\rho(H)/\rho(30\,\mathrm{T})$, measured at 0.5 K for B1 microstructure (red) and A2 single crystal (blue). **b** The oscillatory part of the magnetoresistance $\Delta\rho_{\mathrm{osci}}$ obtained by subtracting a polynominal background from $\rho(H)$. © 2021 The author

increasing H due to the suppression of Kondo gap. The MR displays clear wiggle-like features from about 33 T to 42 T in A2 single crystal, while no discernible wiggle features are observed in B1 microstructure. This can be seen clearly in Fig. 4.5b, which shows oscillatory part of the resistivity $\Delta\rho_{\mathrm{osci}}$ obtained after subtracting a polynominal background from $\rho(H)$. For A2 single crystal, the valleys in $\Delta\rho_{\mathrm{osci}}$ are approximately uniformly spaced as a function of $1/H$. On the other hand, no periodic oscillation with respect to $1/H$ is observed up to the insulator-metal transition field of $\mu_0 H_{\mathrm{IM}} \sim 47\,\mathrm{T}$ in B1 microstructure where the bulk contribution is almost 1/100 times smaller than that in A2. We note that the surface resistance is almost unchanged in single crystals and microstructures, indicating that the scattering rate is comparable in all the samples. Therefore, the SdH oscillations are expected to be observed in B1 microstructure if SdH oscillations arise from the surface conducting channel. These results provide supporting evidence that the SdH oscillations in YbB_{12} occur in the insulating bulk.

4.5 Conclusion

In summary, we studied transport properties of KI YbB_{12} microstructures fabricated by FIB technique. The low-temperature resistivity plateau decreases with reducing the sample thickness, demonstrating the presence of the surface metallic channels. The MR of microstructures displays a distinct WAL effect, which is consistent with the spin-momentum locking due to the topological protection. We also present SdH effect is not observed in the microstructures, suggesting that unconventional QOs are inherent properties of the insulating bulk state.

References

1. Tsunetsugu H, Sigrist M, Ueda M (1997) Rev Mod Phys 69:809–863
2. Riseborough PS (2000) Adv Phys 49:257–320
3. Li L, Sun K, Kurdak C, Allen JW (2020) Nat Rev Phys 2:463–479
4. Dzero M, Xia J, Galitski V, Coleman P (2016) Annu Rev Condens Matter Phys 7:249
5. Dzero M, Sun K, Galitski V, Coleman P (2010) Phys Rev Lett 104:106408
6. Kim DJ, Xia J, Fisk Z (2014) Nat Mater 13:466
7. Syers P, Kim D, Fuhrer MS, Paglione J (2015) Phys Rev Lett 114:096601
8. Eo YS, Wolgast S, Rakoski A, Mihaliov D, Kang BY, Song MS, Cho BK, Hatnean MC, Balakrishnan G, Fisk Z, Saha SR, Wang X, Paglione J, Kurdak C (2020) Phys Rev B 101:155109
9. Xu N, Biswas PK, Dil JH, Dhaka RS, Landolt G, Muff S, Matt CE, Shi X, Plumb NC, Radović M, Pomjakushina E, Conder K, Amato A, Borisenko SV, Yu R, Weng HM, Fang Z, Dai X, Mesot J, Ding H, Shi M (2014) Nat Commun 5:4566
10. Weng H, Zhao J, Wang Z, Fang Z, Dai X (2014) Phys Rev Lett 112:016403
11. Hagiwara K, Ohtsubo Y, Matsunami M, Ideta SI, Tanaka K, Miyazaki H, Rault JE, Le Févre P, Bertran F, Taleb-Ibrahimi A, Yukawa R, Kobayashi M, Horiba K, Kumigashira H, Sumida K, Okuda T, Iga F, Kimura SI (2016) Nat Commun 7:12690
12. Li G, Xiang Z, Yu F, Asaba T, Lawson B, Cai P, Tinsman C, Berkley A, Wolgast S, Eo YS, Kim DJ, Kurdak C, Allen JW, Sun K, Chen XH, Wang YY, Fisk Z, Li L (2014) Science 346:1208
13. Tan BS, Hsu YT, Zeng B, Ciomaga Hatnean M, Harrison N, Zhu Z, Hartstein M, Kiourlappou M, Srivastava A, Johannes MD, Murphy TP, Park JH, Balicas L, Lonzarich GG, Balakrishnan G, Sebastian SE (2015) Science 349:287
14. Xiang Z, Kasahara Y, Asaba T, Lawson B, Insman C, Chen L, Sugimoto K, Kawaguchi S, Sato Y, Li G, Yao S, Chen YL, Iga F, Singleton J, Matsuda Y, Li L (2018) Science 69:65
15. Sato Y, Xiang Z, Kasahara Y, Taniguchi T, Kasahara S, Chen L, Asaba T, Tinsman C, Murayama H, Tanaka O, Mizukami Y, Shibauchi T, Iga F, Singleton J, Li L, Matsuda Y (2019) Nat Phys 15:954
16. Xiang Z, Chen L, Chen K-W, Tinsman C, Sato Y, Asaba T, Lu H, Kasahara Y, Jaime M, Balakirev F, Iga F, Matsuda Y, Singleton J, Li L (2021) Nat Phys 17:788
17. Sato Y, Xiang Z, Kasahara Y, Kasahara S, Chen L, Tinsman C, Iga F, Singleton J, Nair NL, Maksimovic N, Analytis JG, Li L, Matsuda Y (2021) J Phys D Appl Phys 54:404002
18. Iga F, Shimizu N, Takabatake T (1998) J Magn Magn Mater 177–181:337
19. Wolgast S, Kurdak Ç, Sun K, Allen JW, Kim DJ, Fisk Z (2013) Phys Rev B 88:180405(R)
20. Gabáni S, Pristáš G, Takáčová I, Sluchanko N, Siemensmeyer K, Shitsevalova N, Filipov V, Flachbart K (2015) Solid State Sci 47:17–20
21. Harrison N (2018) Phys Rev Lett 121:026602
22. Shen H, Fu L (2018) Phys Rev Lett 121:026403
23. Wakeham N, Wang YQ, Fisk Z, Ronning F, Thompson JD (2015) Phys Rev B 91:085107
24. De S, Coleman JN (2010) SACS Nano 4(5):2713–2720
25. Ando T, Nakanishi T, Saito R (1998) J Phys Soc Jpn 67:2857
26. Hasan MZ, Kane CL (2010) Rev Mod Phys 82:3046
27. Qi XL, Zhang SC (2011) Rev Mod Phys 83:1057
28. Thomas S, Kim DJ, Chung SB, Grant T, Fisk Z, Xia J (2016) Phys Rev B 94:205114
29. Assaf BA, Cardinal T, Wei P, Katmis F, Moodera JS, Heiman D (2013) Appl Phys Lett 102:012102
30. Sugiyama K, Iga F, Kasaya M, Kasuya T, Date M (1988) J Phys Soc Jpn 57:3946
31. Shahrokhvand M, Pezzini S, Van Delft MR, Zeitler U, Hussey NE, Wiedmann S (2017) Phys Rev B 96:205125
32. Hikami S, Larkin AI, Nagaoka Y (1980) Prog Theor Phys 63:707
33. Pavlosiuk O, Kaczorowski D, Fabreges X, Gukasov A, Wisniewski P (2016) Sci Rep 6:18797
34. Sasmal S, Mondal R, Kulkarni R, Thamizhavel A, Singh B (2020) J Phys Condens Matter 32:335701

Chapter 5
Charge-Neutral Fermions in YbB_{12}

5.1 Introduction

Kondo insulators (KIs) have attracted renewed interest owing to the introduction of the concept of topology. Moreover, a series of recent detailed experiments have revealed another aspect of exotic properties in this strongly correlated insulator: unconventional quantum oscillations (QOs). The QOs observed in SmB_6 and YbB_{12} sparked an intensive debate, but their origin remains unclear. As these observations generally indicate the presence of a Fermi surface, it is interesting to see if another experimental technique can resolve it. If they host a Fermi surface with charge neutral fermions (CNFs), as some theories proposed, their zero-energy excitations might contribute to the heat transport, resulting in a linear temperature dependence of the specific heat C and thermal conductivity κ_{xx}. While the finite $\gamma = C/T(T \to 0)$ results from both localized and itinerant excitations, a finite residual thermal conductivity $\kappa_{xx}^0/T = \kappa_{xx}/T(T \to 0)$ is exclusively derived from itinerant excitations [1]. As thermal conductivity is free from local excitations such as the Schottky anomaly, it can provide the most direct and compelling evidence to determine whether the fermionic excitations have an itinerant character. Furthermore, if these neutral fermions are responsible for the QOs at higher fields, the possible Landau quantization of CNFs might contribute to the finite thermal Hall signal in analogy with conventional electrons. Motivated by these speculations, we studied the low-energy excitations in the KI YbB_{12}.

It has been reported that SmB_6 shows a finite γ [2–4]. Figure 5.1a shows the temperature dependence of C/T for a series of single crystals of SmB_6 [3]. C/T shows a strong sample dependence, and the magnetic doping dramatically enhances the low-T upturn. Although it was reported that the latest and cleanest sample grown using the floating zone method shows a finite $\gamma \approx 4$ mJ/molK^2, there is ongoing debate on whether finite γ is intrinsic to the insulating bulk, intrinsic to the in-gap state, or an extrinsic impurity contribution. The thermal conductivity of SmB_6 was also reported by several groups [4–6], and these results are shown in Figs. 5.1b–d. Although κ_{xx}^0/T

Y. Sato, *Quantum Oscillations and Charge-Neutral Fermions in Topological Kondo Insulator YbB_{12}*, Springer Theses,
https://doi.org/10.1007/978-981-16-5677-4_5

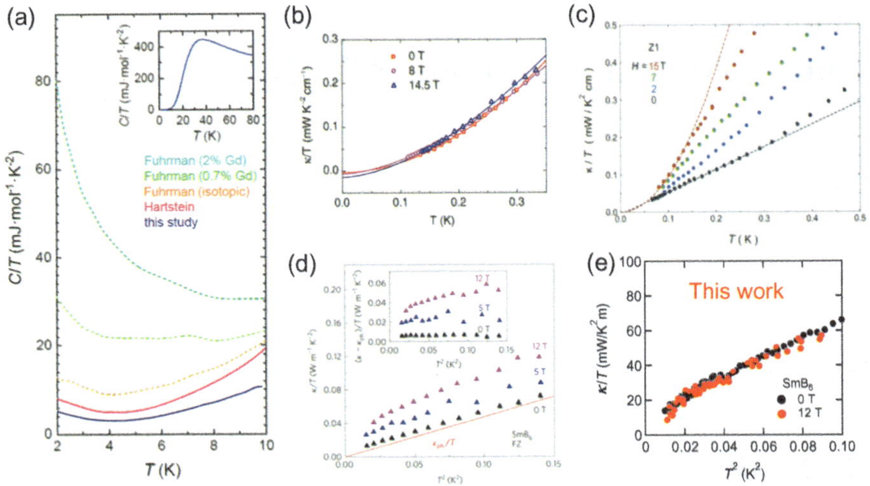

Fig. 5.1 **a** Low-temperature specific heat in a series of single crystals of SmB_6. Reprinted from [3] by The Author licensed under CC BY 4.0. **b** Temperature dependence of thermal conductivity divided by T, κ/T in SmB_6 single crystal grown by the flux method. Reprinted figures with permission from [5] © 2016 by the American Physical Society. **c** κ/T vs. T in floating zone sample with various fields. Reprinted figures with permission from [6] © 2018 by the American Physical Society. **d** κ/T versus T^2 in floating zone sample with various fields. Reprinted by permission from Springer Nature: Nature Physics [4], © (Copyright Springer Nature Singapore Pte Ltd. 2018) **e** Thermal conductivity of SmB_6 obtained in this work. © 2021 The author.

was reported to be close to zero in these references, the effect of magnetic field on κ_{xx} has been reported differently among the groups. While reference [5] reports a field-independent thermal conductivity, a significant magneto-thermal conductivity is reported in reference [6], which is interpreted in terms of conventional phonon scattering contributions. We also measured the low-temperature thermal conductivity of SmB_6 synthesized using the floating zone method, as shown in Fig. 5.1e. Our data at 0 T and 12 T do not show a significant field dependence, and in both sets of data, κ_{xx}/T becomes vanishingly small on approaching zero temperature. These results indicate the absence of itinerant neutral fermionic excitations in SmB_6. Although inelastic neutron scattering experiments reveal distinct excitation modes within the hybridization gap [7], it is not evident whether the excitations are charge-neutral. Thus, the presence of intrinsic non-trivial itinerant quasiparticles within the gap in SmB_6 remains a controversial issue.

In the remainder of this chapter, we present measurements of the thermal transport properties of another candidate TKI, YbB_{12}, to investigate the low-energy excitations [8]. Although SmB_6 and YbB_{12} show quite similar electronic properties, including the unconventional QOs, there seem to be several salient differences, such as in the presence of the SdH effect and the effective mass m^* determined by the QOs. Therefore, systematic studies covering a wide range of materials may be key to understanding the nature of the exotic physics found in TKIs.

5.2 Experimental

5.2.1 *Specific Heat*

Specific heat is one of the most fundamental physical properties of matter, as it is a bulk thermodynamic quantity related to the free energy of a given system. As the measurement of specific heat provides pivotal information on the electronic structure of a many-body system, it has been employed as a powerful tool for studying magnetism, superconductors, correlated insulators, and so on. The specific heat C is defined as the ratio of the change in the heat absorbed by the system δQ to the change in temperature δT:

$$C = T\left(\frac{\partial S}{\partial T}\right)_p = \lim_{\delta T \to 0} \frac{\delta Q}{\delta T}. \tag{5.1}$$

A variety of experimental techniques have been developed to measure specific heat, such as adiabatic calorimetry, the continuous heating method, and AC calorimetry. In this study, we developed a setup based on the long-relaxation method, which has an overwhelming advantage in the estimation of specific heat of tiny single crystals with $\approx 1\,\mu$g in weight [9, 10].

The principle of the long-relaxation method is as follows. A schematic of the measurement is shown in Fig. 5.2a. The sample is mounted on a bare thermometer, which can also be used as a heater by applying excitation currents, producing Joule heat with power $P(T)$. The thermometer is weakly connected to a heat bath through a connection with a small thermal conductivity of $\kappa(T)$. In the long-relaxation method, a relatively large amount of heat is injected into the sample, and the time evolution of the sample temperature is recorded. Because of the large heat injection, the change in C and κ with respect to T cannot be negligible. Taking the derivative of Eq. 5.1 with respect to time, we obtain

$$\frac{\mathrm{d}Q}{\mathrm{d}t} = C(T)\frac{\mathrm{d}T}{\mathrm{d}t}, \tag{5.2}$$

which can be expressed in the thermal equilibrium equation as

$$P(T) - \int_{T_0}^{T} \kappa(T')\mathrm{d}T' = C(T)\frac{\mathrm{d}T}{\mathrm{d}t}. \tag{5.3}$$

In this method, we control $P(T)$ in the manner shown in Fig. 5.1b: A higher power $P_{\mathrm{High}}(t)$ and a lower power $P_{\mathrm{Low}}(t)$ are fed into the sample within a constant time interval Δt. The temperature evolution with respect to time with the injection of a higher and lower power is then expressed as $T_{\mathrm{High}}(t)$ and $T_{\mathrm{Low}}(t)$, respectively. According to Eq. 5.3, these two relaxation processes can be written as

$$C(T)\frac{dT_{\mathrm{High}}}{dt} = P_{\mathrm{High}}(T) - \int_{T_0}^{T} \kappa(T')dT' \tag{5.4}$$

$$C(T)\frac{dT_{\mathrm{Low}}}{dt} = P_{\mathrm{Low}}(T) - \int_{T_0}^{T} \kappa(T')dT'. \tag{5.5}$$

This pair of equations can then be combined to obtain

$$C(T) = \frac{P_{\mathrm{High}}(T) - P_{\mathrm{Low}}(T)}{\left[dT_{\mathrm{High}}/dt - dT_{\mathrm{Low}}/dt\right]_T}. \tag{5.6}$$

Thus, by monitoring the time evolution of the relaxation curves with two different powers, the specific heat can be calculated using the above formula. The advantage of this technique is that we need not know $\kappa(T)$ accurately, because it will be canceled out as long as the base temperature T_0 is not influenced by the heat power. Another merit is that we can collect a set of data on $C(T)$ during just a pair of relaxation curves in a temperature range of $T_0 < T < T_{\mathrm{max}}$.

In DC measurements, the voltage induced by the thermoelectric effect cannot be negligible. This voltage can be canceled out by repeating the set of relaxations with positive and negative currents. One sequence for the measurement, therefore, consists of four sets of currents, $+I_{\mathrm{High}}$, $+I_{\mathrm{Low}}$, $-I_{\mathrm{High}}$, and $-I_{\mathrm{Low}}$, and the reading of the corresponding voltages, $V_{+\mathrm{High}}$, $V_{+\mathrm{Low}}$, $V_{-\mathrm{High}}$, and $V_{-\mathrm{Low}}$. The thermoelectric voltage can then be canceled as

$$V_{\mathrm{High}} = \frac{V_{+\mathrm{High}} + V_{-\mathrm{High}}}{2} \tag{5.7}$$

$$V_{\mathrm{Low}} = \frac{V_{+\mathrm{Low}} + V_{-\mathrm{Low}}}{2}. \tag{5.8}$$

The corresponding resistances of the sample thermometer for both relaxation curves $R_{\mathrm{High}}(t) = V_{\mathrm{High}}/I_{\mathrm{High}}$ and $R_{\mathrm{Low}}(t) = V_{\mathrm{Low}}/I_{\mathrm{Low}}$ are converted to temperature using the calibration T-R table, which is already determined. Figure 5.2c schematically shows the procedure of data analysis. First, the relaxation curves are measured repeatedly (typically 5–100 times according to the signal-to-noise ratio (S/N) under the given conditions), following which they are averaged. Next, we take the derivative of T with respect to t. While T is recorded within the same time interval Δt, the resulting $dT/dt(T)$ is not obtained at regular intervals with respect to T. Therefore, one must perform linear interpolation to obtain a pair of dT_{High}/dt and dT_{Low}/dt at the same T. The temperature dependence of $P(T)$ is also taken into account because it changes according to $P(T) = I^2 \cdot R(T)$, where the resistance of the heater $R(T)$ is identical to that of the thermometer. Finally, we obtain a set of points $C(T)/T$ by using Eq. 5.6.

Figure 5.3a–c shows images and a schematic of the experimental setup. A Cernox 1030 BR chip was used as a thermometer, sample stage, and heater (all of these terms

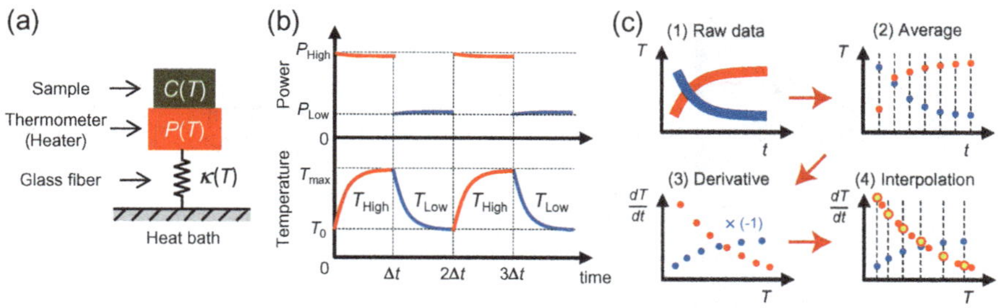

Fig. 5.2 **a** Schematic of the setup for specific heat measurement using the long-relaxation method. **b** Time evolution of the temperature response to applied power. **c** Illustrations of the procedure of data analysis. © 2021 The author.

refer to the chip). In an actual measurements, the addenda contribution, which is the background specific heat observed even without the sample, should be minimized to achieve a good S/N. For this purpose, the bare thermometer chip was employed in the simplest possible form. A gold surface, which was coated on the Cernox chip, was removed by polishing with sandpaper, leading to decreased addenda and better thermal connection to samples. To suspend the thermometer in vacuum with a weak thermal connection to the heat bath, we used a pair of glass fibers of $\phi = 30\,\mu$m, which were coated by gold to form electrical contacts. The resistance of the glass fiber was $\sim$ 80 Ω. The contact to the thermometer and heat bath was made by silver paint. Figure 5.3b shows an image taken after mounting a sample. The sample was fixed on the stage by applying a tiny amount of high-vacuum grease. The copper base was thermally connected to the ^{3}He pot of an Oxford Heliox. The circuit used in the long-relaxation method is illustrated in Fig. 5.3c. The resistance of the sample thermometer was measured using the conventional four-terminal configuration. The excitation was applied using a Keithley 6221 current source, producing simultaneous heat pulses to the sample. The voltage was then monitored using Keysight 3458A voltmeter in the fast reading mode. Furthermore, resistors $R_I = 200$ kΩ and $R_V \sim$ 0–30 kΩ were used to remove the noise and optimize S/N according to each experimental condition. The filter boxes consist of low-pass filters with capacities of 22,000 pF and 100 pF at room temperature and cryostat temperatures, respectively.

In Fig. 5.4a, we show the temperature dependence of the specific heat of addenda C_{Add} and total specific heat C_{Tot} including that of the YbB_{12} sample, as shown by the red and black curves, respectively. For comparison, C_{Add} obtained using the quasi-adiabatic method, which has been employed as a conventional technique in the lab, is shown by the blue curve. C_{Add} measured using the quasi-adiabatic method is almost twice as large as C_{Tot}. On the other hand, C_{Add} obtained using the newly developed long-relaxation method is much smaller, and it becomes $\sim 1/10$ times smaller near the lowest temperature of 0.5 K. It is also important to consider the field dependence of C_{Add} as it cannot be negligible when the sample signal is comparable to that of addenda. In fact, a significant field dependence is observed, as displayed in Fig. 5.4b. Therefore, each time we change the sample, C_{Add} was re-calibrated because even a minute amount of glue can significantly affect the results.

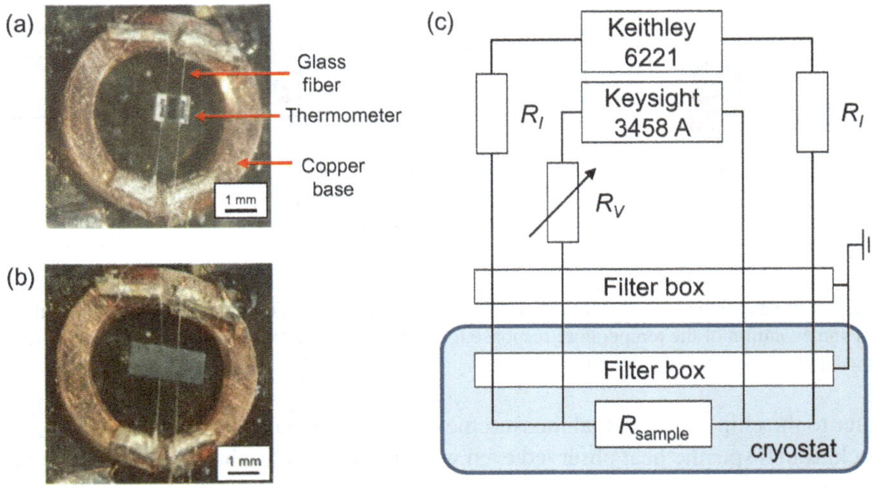

Fig. 5.3 **a**, **b** Images of the setup for the specific heat measurement. The thermometer is connected to a copper heat bath by glass fibers with $\phi = 30\ \mu$m **a** without a sample and **b** with a YbB_{12} sample. **c** Schematic of the circuit used in the measurements. © 2021 The author

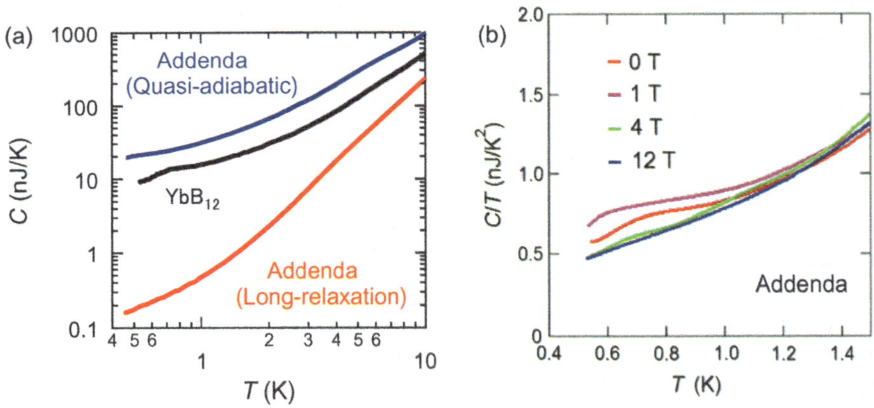

Fig. 5.4 **a** Log–log plot of the measured total specific heat C without (the red curve) and with a YbB_{12} sample (the black curve) as a function of T. For comparison, the addenda data obtained using the quasi-adiabatic method (the blue curve) are also displayed. **b** Temperature dependence of the addenda contribution measured using the long-relaxation method with various fields. © 2021 The author.

5.2.2 Thermal Transport

In addition to the electrical transport, thermal conductivity measurements provide pivotal information on both charged and neutral quasiparticle excitations of the underlying system. We will describe the principles and experimental details below. Passing a heat current through materials induces a thermal gradient. In the steady state, this response can be written as

$$q_i = -\kappa_{ij} \partial_j T, \tag{5.9}$$

where q_i $(i = x, y, z)$ is the thermal current density along the i-direction, κ_{ij} is the thermal conductivity tensor, and $\partial_j T$ is the thermal gradient along the j-direction. Assuming that the heat flows only in the xy-plane for simplicity, this relation can be expressed as

$$\begin{pmatrix} q_x \\ q_y \end{pmatrix} = \begin{pmatrix} \kappa_{xx} & \kappa_{xy} \\ -\kappa_{xy} & \kappa_{xx} \end{pmatrix} \begin{pmatrix} \partial_x T \\ \partial_y T \end{pmatrix}. \tag{5.10}$$

A plate-like sample is desirable to well define a uniform thermal current and thermal gradient. In a typical experimental setup, one side of the sample is attached to the heat bath, while the other side is connected to the heater, as shown in Fig. 5.5a. In this setup, $q_x = Q/wt$, $q_y = 0$, $-\partial_x T = \Delta T_x/\ell$, and $-\partial_y T = \Delta T_y/w'$. Here, Q is the power released from the heater, and ΔT_x and ΔT_y are the longitudinal and transverse thermal difference given by $\Delta T_x = T_{\mathrm{H}} - T_{\mathrm{L}}$ and $\Delta T_y = T_{\mathrm{L}} - T_{\mathrm{L'}}$, respectively. ℓ, t, and w are dimensional factors: length, thickness, and width, respectively. It is worth noting that there are two different definitions for width: w is measured from one sample edge to the other, while w' is defined as the distance between two thermal contacts from L to L'. Ignoring $O(\Delta T_y)^2$, each component of the thermal conductivity tensor is finally given by

$$\begin{aligned} \kappa_{xx} &= \frac{Q}{\Delta T_x} \frac{\ell}{wt}, \\ \kappa_{xy} &= \left(\frac{\Delta T_y}{Q}\right) \left(\frac{\Delta T_x}{Q}\right)^2 \frac{\ell^2}{ww't} \\ &= \left(\frac{\Delta T_y}{Q}\right) \frac{wt}{w'} \kappa_{xx}^2. \end{aligned} \tag{5.11}$$

In practice, the thermal contacts are inevitably misaligned. Therefore, the transverse response can be contaminated by the longitudinal signal, and vice versa. One can cancel this effect by inverting the sign of the field direction, as the longitudinal and transverse responses are symmetric and antisymmetric, respectively, with respect to the external magnetic field:

$$\Delta T_x(H) = \Delta T_x(-H), \tag{5.12}$$

$$\Delta T_y(H) = -\Delta T_y(-H). \tag{5.13}$$

Therefore, the misalignment effect can be canceled out as

$$\Delta T_x(|H|) = \frac{\Delta T_x(H) + \Delta T_x(-H)}{2} \tag{5.14}$$

$$\Delta T_y(|H|) = \frac{\Delta T_y(H) - \Delta T_y(-H)}{2}. \tag{5.15}$$

Although this cancelation works in principle, it is still important to reduce the misalignment to improve S/N, especially when the κ_{xy} of the sample is small. The misalignment angle can be estimated as $\Delta T_y(0)/\Delta T_x(0)$. In the following thermal Hall conductivity experiments on YbB_{12}, we achieved a misalignment of less than 1%.

The sequence of the measurements is as follows. We stabilize the base temperature T_B and record the resistance of the thermometers R_S without applying power. These sets of data were used for the T-R calibration table of each thermometer. We then applied a heat current to the sample and recorded the thermal gradient to calculate κ_{ij} using Eq. 5.11. If we apply too much power to the sample, the linear response of the thermal gradient ($\Delta T \propto Q$) cannot be reliably evaluated. Therefore, we applied a moderate power so that $\Delta T_x/T_B \sim$ 0.5–2 % and ensured that the thermal gradient response is always Q-linear.

Figure 5.5b shows an image of the actual experimental setup. To apply a thermal current along the x-direction uniformly, we used a thin silver plate as a thermal contact to the heater. In addition, if the κ_{xy} of the sample is too small, there may be a possible major contamination of ΔT_y induced by the κ_{xy} of the copper heat bath. Thus, we utilize lithium fluoride (LiF) as an electrically insulating heat bath. In fact, LiF has a large activation gap $\sim$ of 15 eV [11] and exhibits a vanishingly small ΔT_y, which can be negligibly small in the same setup [12]. The sample, LiF (the transparent material shown in Fig. 5.5b), and copper base were connected with high-vacuum grease, whereas the thermal contacts to the sample were made using silver paint.

We schematically show our sample cell in Fig. 5.6a. To measure the sample temperature precisely, we employed a RuO_2 chip resistor as a sample thermometer. As one has to place the sample and thermometer in a practical vacuum environment, it is desirable to minimize possible thermal connections to the bath (i.e., thermal ground). For this purpose, less conductive wires of twisted manganin with $\phi = 30\,\mu$m and $R_{mang} \sim 200\ \Omega$ were used to make electrical contacts to the thermometers, which are rolled into coils to increase the resistance in the limited space of our sample cell. The thermometers are placed on a piece of polyimide tubes (DuPont, kapton tubes), which are attached to a fiber-reinforced plastics (FRP) frame. To arrange the three RuO_2 chips and one heater independently, the sample cell consists of four sets of the same structure, including a manganin coil, thermometer/heater, and prop kapton

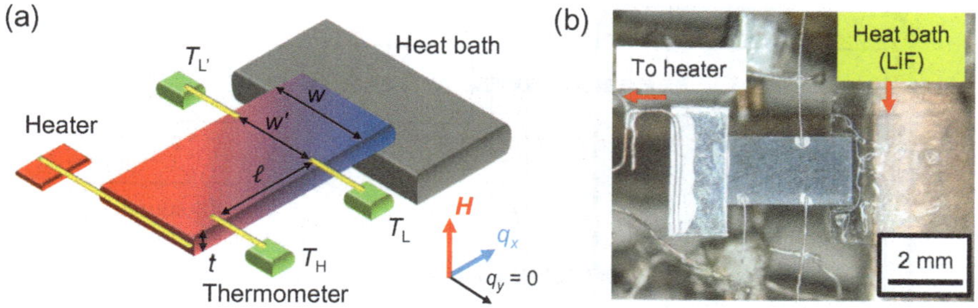

Fig. 5.5 **a** Schematic of thermal conductivity measurement using the steady-state method. **b** Image of the actual setup. The YbB_{12} sample (the black rectangle in the middle) is placed on the LiF heat bath (the transparent object). A heat current is passed through the silver plate (the silver rectangle on the left), which is connected to the heater. © 2021 The author.

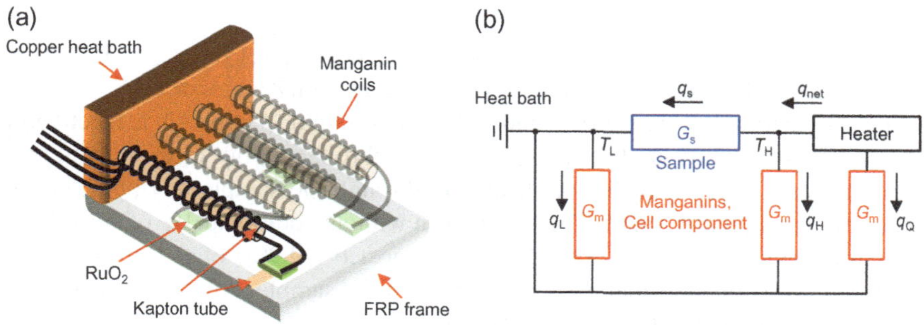

Fig. 5.6 **a** Schematic of the sample cell for thermal conductivity measurement. The sample is attached to the bottom surface of the copper heat bath. A sample and the contacts between thermometers are excluded in the depiction for visibility. **b** Thermal circuit for our setup describing possible heat leaks through the manganin coils and cell components such as the FRP frame. © 2021 The author.

tube. For visibility, we only show one of the above structures clearly in Fig. 5.6a; the others are transparent.

In an ideal setup, one can simply assume that the heat current released from the heater passes solely through the sample and heat ground. However, in practice, heat leakage is inevitable through the components of the sample cell, such as the manganin wires and frame. First, if the resistance of the heater is comparable to that of the manganin wire, the Joule heating in the wire cannot be negligible. To avoid this, we used $R_Q = 10\ \mathrm{k\Omega}$, which is more than 100 times larger than R_{mang}. In addition, as shown in Fig. 5.6b, there are several possible heat-leak paths through the thermal conductance G_{m}. If this heat leak is comparable to that from the sample thermal conductance G_{s}, one cannot evaluate the heat current q_{s} through the sample reliably. According to reference [13], the κ_{xx} of manganin is approximately 0.05 and 1 mW/cmK at 0.1 K and 1 K, respectively. This leads to $G_{\mathrm{m}}(0.1\ \mathrm{K}) \sim 7 \times 10^{-11}$ W/K and $G_{\mathrm{m}}(1\ \mathrm{K}) \sim 1.4 \times 10^{-9}$ W/K for our coils. On the other hand, the G_{s} of a

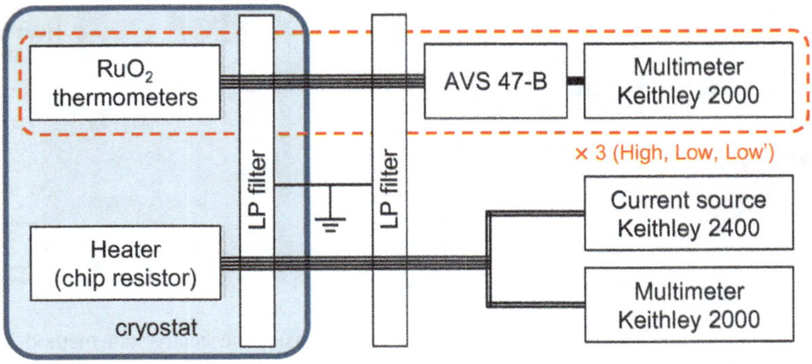

Fig. 5.7 Schematic of the circuit used for the thermal conductivity measurement. © 2021 The author.

typical YbB_{12} crystal we measured can be estimated as $G_s(0.1\ \mathrm{K}) \sim 5 \times 10^{-8}$ W/K and $G_s(1\ \mathrm{K}) \sim 1 \times 10^{-5}$ W/K. Here, we used a dimensional factor of $wt/\ell \sim 50\ \mu$m for the calculation ($wt/\ell \sim 37\mu$m and $\sim 63\ \mu$m in sample #1 and #3, respectively). The above estimations imply that $G_s/G_m > 10^3$ throughout our measurement temperature range, indicating that the heat leak through the sample thermometer and manganin is negligibly small. Therefore, even with an additional 1% of Joule heating within the manganin wire of the heater, most (~99.9%) of the power would contribute to the net thermal current q_{net} toward the sample. Although an accurate estimation of the heat leak through other components such as the frame is difficult, we ensured that without the sample, none of the thermometers responded to the typical power we used for the measurements.

Figure 5.7 schematically shows the circuit for the thermal transport measurements. To simultaneously monitor the temperature of three different points, High, Low, and Low', we prepared three sets of AVS 47-B resistance bridges. The measured resistance is converted through the DC output line to a Keithley 2000 multimeter. Regarding the magnitude of heat current, an excitation current of I_Q was applied from the Keithley 2400 current source. As the T- and H-dependence of the resistance in the heater R_Q must be considered, the voltage V_Q was also recorded using another Keithley 2000 multimeter. The heat power is given from the observed V_Q as $Q = I_Q \cdot V_Q$. Low-pass (LP) filter boxes were used to cut external noise as well as for the specific heat measurements. We also used a 12–14 T solenoidal superconducting magnet (Oxford Instruments) to apply magnetic fields and a gas handing system with a ^{3}He-^{4}He dilution refrigerator (Cryoconcept, DR-JT-S-200-10), the lowest temperature of the mixing chamber of which is ~40 mK.

5.3 Results and Discussion

5.3.1 Specific Heat

Figure 5.8a shows the temperature dependence of the specific heat divided by the temperature (C/T) of #2 and #3 (the same single crystals as presented in Chap. 3) in zero field as blue filled and open circles, respectively. The overall temperature dependence has several characteristic structures: a slight upturn below ~1 K, a broad hump-like anomaly around 6 K, and a steep growth above ~10 K. It is notable that despite the upturn at low temperature, C/T does not seem to approach zero as $T \to 0$: The extrapolation from above 1 K clearly has a finite intercept. This indicates that the specific heat of YbB_{12} has a linear temperature dependence, which can be expressed as $C_{qp} = \gamma T$, suggesting that the ground state has gapless fermionic excitations similar to those of ordinal metals. To quantitatively evaluate the T-linear term, we estimated other contributions as follows. First, the slight upturn below 1 K is attributed to a Schottky contribution, C_{Sc}/T. The low-temperature enhancement of C/T is well fitted by a three-level Schottky model, as shown by the green solid and dotted curves in Fig. 5.8a. Second, the hump anomaly and steep growth above ~ 10 K can be described by low-energy Einstein optical phonon modes. These anomalies have also been reported in the isostructual metallic compounds LuB_{12} and YB_{12} [14]. The MB_{12} (M is a rare-earth element) structure can be regarded as free oscillators of rare-earth atoms in rigid cavities of B_{24} cuboctahedrons, as shown in Fig. 5.9a. Figure 5.9b shows the specific heat of the isostructual LuB_{12} and YB_{12}. Similar to our data for YbB_{12}, both compounds show a steep upturn in C above ~10 K. In addition, inelastic neutron scattering experiments of YbB_{12} reveal 3D phonon dispersions [15]. Therefore, we attribute both the acoustic phonon ($\propto T^3$) and two optical phonon modes to C_{ph}, which is described by the solid and dotted black lines in Fig. 5.8a. Owing to the high Debye temperature, the acoustic phonon contribution to the total heat capacity is very small. On the other hand, the optical phonon contributions are slightly sample dependent, but the other components are almost the same in #2 and #3.

The low-temperature specific heat of YbB_{12}, therefore, can be written as a sum of the phonon, quasiparticle, and Schottky contributions as

$$C = C_{qp} + C_{ph} + C_{Sc}, \tag{5.16}$$

where

$$C_{qp} = \gamma T, \tag{5.17}$$

$$C_{ph} = \beta T^3 + \frac{A}{T}\left(\frac{\Theta_E}{T}\right)^2 \frac{\exp(\Theta_E/T)}{[\exp(\Theta_E/T) - 1]^2}, \tag{5.18}$$

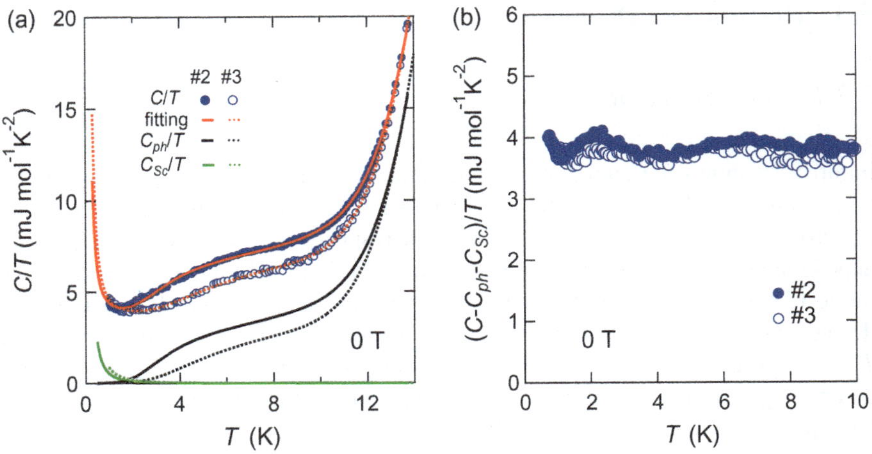

Fig. 5.8 **a** Specific heat data at zero field in #2 and #3 (the blue filled and open circles, respectively). The red, black, and green curves represent the total fitting results, phonon contribution, and Schottky contribution, respectively. **b** Temperature dependence of C/T subtracted by the phonon and Schottky contributions. © 2021 The author

$$C_{\mathrm{Sc}} = \frac{B}{T^2}\left[\frac{\sum_{i=0}^{n-1}\epsilon_i^2\exp(-\epsilon_i/k_{\mathrm{B}}T)}{\sum_{i=0}^{n-1}\exp(-\epsilon_i/k_{\mathrm{B}}T)} - \left(\frac{\sum_{i=0}^{n-1}\epsilon_i\exp(-\epsilon_i/k_{\mathrm{B}}T)}{\sum_{i=0}^{n-1}\exp(-\epsilon_i/k_{\mathrm{B}}T)}\right)^2\right]. \quad (5.19)$$

Here, Θ_{E} is the Einstein phonon temperature, and ϵ_i (i = 1, 2, 3) is the energy levels yielding the Schottky contribution. As shown in Fig. 5.8b, C_{qp}/T obtained by subtracting C_{ph} and C_{Sc} from the total C is consistent between #2 and #3. The calculated γ is ≈ 3.8 mJ/molK^2 in zero field, which is comparable to the values in isostructual metallic LuB_{12} ($\gamma \approx 4.2$ mJ/molK^2 [14]) and other conventional metals. We stress that the volume fraction of the impurity phases of our single crystals is estimated to be much less than 1% from high-resolution synchrotron X-ray diffraction measurements, as we discussed in Sect. 3.2. Therefore, we can rule out the possibility that the finite γ arises from impurity phases.

Next, we discuss the field dependence of the specific heat. Figures 5.10a–i show the temperature dependence of C/T below 5 K in various magnetic fields up to 12 T. The anomalies mentioned above, i.e., the low-temperature upturn and hump behavior, appear to be field dependent. At 4 T, for example, C/T decreases as T decreases with a downward curvature below 2 K. At 8 and 12 T, on the other hand, C/T decreases almost linearly against T with a steeper slope than that of the zero-field data. This low-temperature behavior may be attributed to the coupling between the magnetic field and the optical phonon modes. The field dependence of C/T at the lowest temperature $T = 1$ K is displayed in Fig. 5.11a. The low-field anomaly around 2 T is mainly attributed to the significant change in the Schottky contribution. We also performed the fitting of C/T in fields by using the same model, Eq. 5.16, and the resulting parameters are displayed in Figs. 5.11b–d. Here, Δ_i is the energy splitting

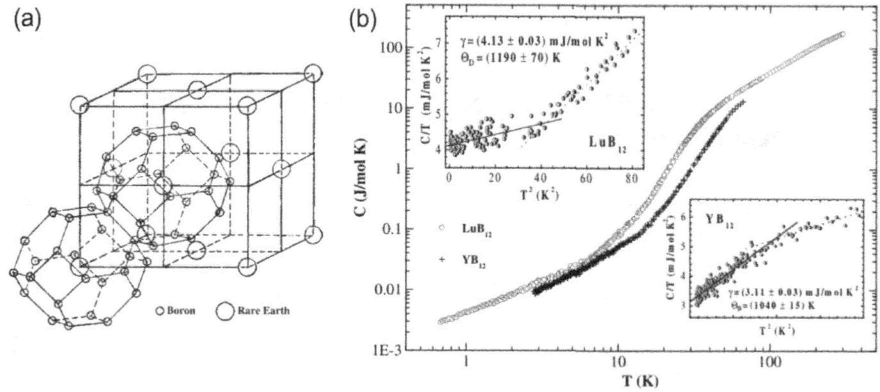

Fig. 5.9 **a** Crystalline structure of MB_{12}. The cluster of boron forms Fedorov's cubooctahedra. **b** Specific heat of the isostructual compounds LuB_{12} and YB_{12} [14].

in the three-level model. As the field increases, γ is slightly reduced, while the Schottky contributions are substantially affected. Although the exact estimation of γ is difficult because of the underlying large Schottky and phonon contributions, the present results imply that the gapless fermionic excitations are affected by external fields.

5.3.2 *Thermal Conductivity*

We now turn to the thermal conductivity, which shows the itinerant aspect of the neutral excitations. The red and blue circles in Fig. 5.12a represent the T^2 dependence of κ_{xx}/T in zero field for #1 and #3, respectively. As the thermal conductivity does not contain the localized Schottky contribution, κ_{xx} can be described as a sum of the itinerant quasiparticle and phonon contributions: $\kappa_{xx} = \kappa_{xx}^{\mathrm{qp}} + \kappa_{xx}^{\mathrm{ph}}$. To use κ_{xx} as a probe of itinerant quasiparticles, $\kappa_{xx}^{\mathrm{ph}}$ must be extracted reliably. The solid lines in Fig. 5.12a are obtained by fitting low-temperature data as follows:

$$\kappa_{xx}/T = \kappa_{xx}^{0}/T + AT^2. \tag{5.20}$$

The AT^2 term here is attributable to phonons for the following reasons. The optical phonon modes are negligible at temperatures well below the Einstein temperature (which is 16–24 K in our samples) because of the small population of optical phonons and the low phonon group velocity. Consequently, acoustic phonons are the only carriers of heat at low temperature, and the phonon thermal conductivity is given by

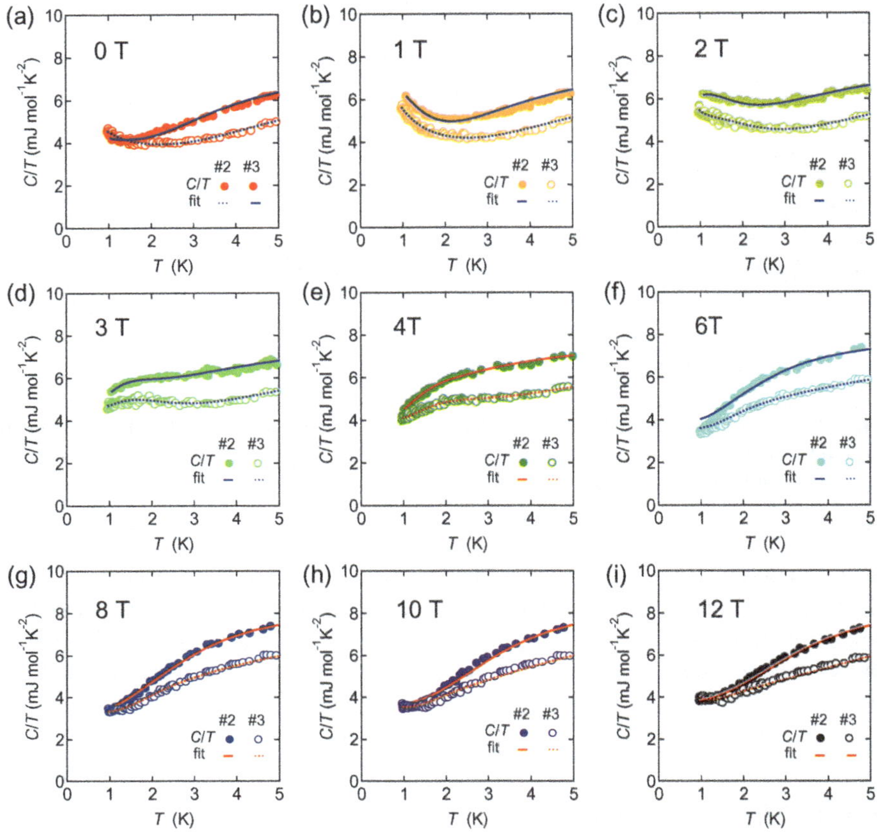

Fig. 5.10 a–i Temperature dependence of C/T in #2 and #3 in various fields up to 12 T. © 2021 The author

$$\kappa_{xx}^{\rm ph} = \frac{1}{3}\beta T^3 v_{\rm ph}\ell_{\rm ph}, \tag{5.21}$$

where $v_{\rm ph}$ and $\ell_{\rm ph}$ are the sound velocity and mean free path of acoustic phonons, respectively. We compare $\ell_{\rm ph}$ and the effective diameter of the sample $d_{\rm eff} = 2\sqrt{wt/\pi}$ (w and t are the width and thickness of the crystal, respectively); $d_{\rm eff}$ = 0.58 mm and 0.37 mm for #1 and #3, respectively. Using β = 0.026 and 0.017 mJ/molK^4 for #1 and #3, respectively, from the specific heat measurements shown in Fig. 5.8a, as well as $v_{\rm ph}$= 9.6 × 10^3 m/s for LuB_{12} [16], we find that $\ell \approx d_{\rm eff}$ at ∼ 0.5 K for #1 and ∼ 0.6 K for #3. These are close to the temperatures below which κ_{xx}/T shows T^2 dependence, as shown in Fig. 5.12a, supporting the above estimation. These results suggest that at sufficiently low temperatures, $\ell_{\rm ph}$ is limited by the crystal size; that is, the samples are in the boundary scattering regime where $\kappa_{xx}^{\rm ph}/T \propto T^2$. The fact that the systems are in this regime is also supported by the A values of #1 and #3. In the boundary scattering regime, A is proportional to $\beta d_{\rm eff}$.

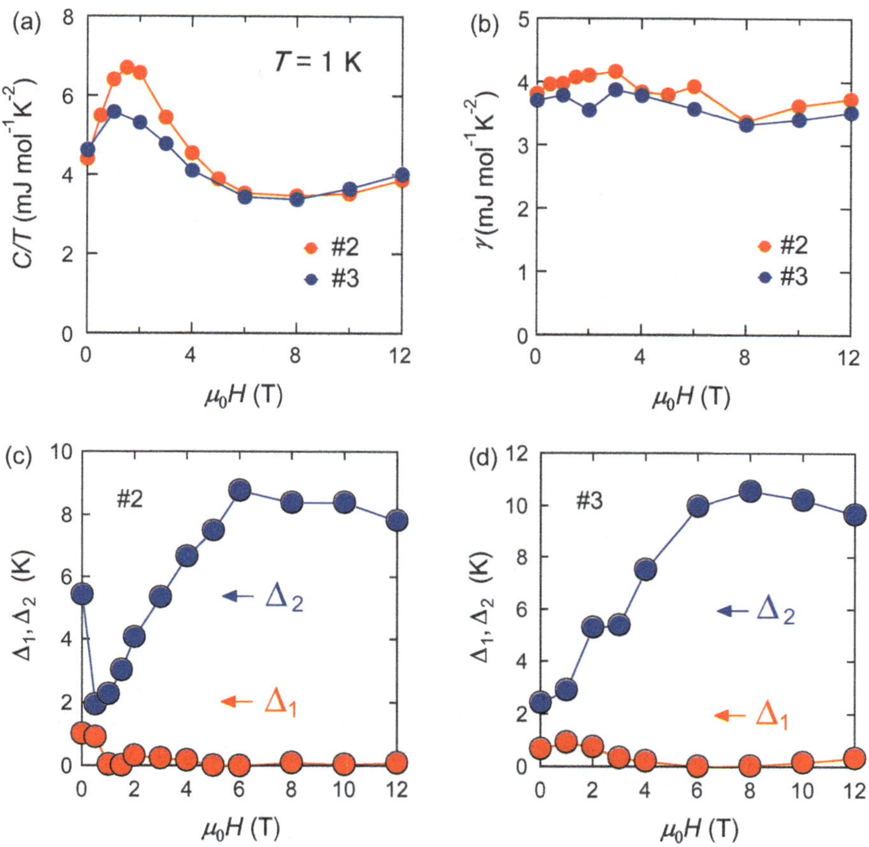

Fig. 5.11 **a** Field dependence of C/T in #2 and #3 at $T = 1$ K. **b** Field dependence of the calculated γ. **c**, **d** Energy splitting of the Schottky specific heat obtained in sample #2 and #3 at different fields. © 2021 The author.

The ratio of the A values between #1 and #3, as determined from the T dependence of κ_{xx}/T, is ~ 2.6. This value is close to the ratio (~ 2.5) of $\beta d_{\rm eff}$ of the two crystals, indicating the proportionality of A and $\beta d_{\rm eff}$.

Figure 5.12b shows the low-temperature thermal conductivity data, where the phonons are in the boundary scattering regime. κ_{xx}/T extrapolated to zero temperature yields definite non-zero intercepts in both crystals: $\kappa^0_{xx}/T \neq 0$. Thus, our results provide evidence of a finite residual linear term in $\kappa^{\rm qp}_{xx}$ (that is, the presence of itinerant gapless fermionic excitations). We stress that the finite κ^0_{xx}/T is not caused by phonons as discussed above. It is known that localized vibrational modes, such as tunneling states in amorphous solids, can contribute to a finite γ [17]. However, such excitations are localized and do not carry heat, resulting in the absence of κ^0_{xx}/T. Moreover, these vibrational modes may act as scattering centers for phonons. This yields a T^{-1} dependence of $\ell_{\rm ph}$, leading to $\kappa_{xx} \propto T^2$, in contrast to the observed

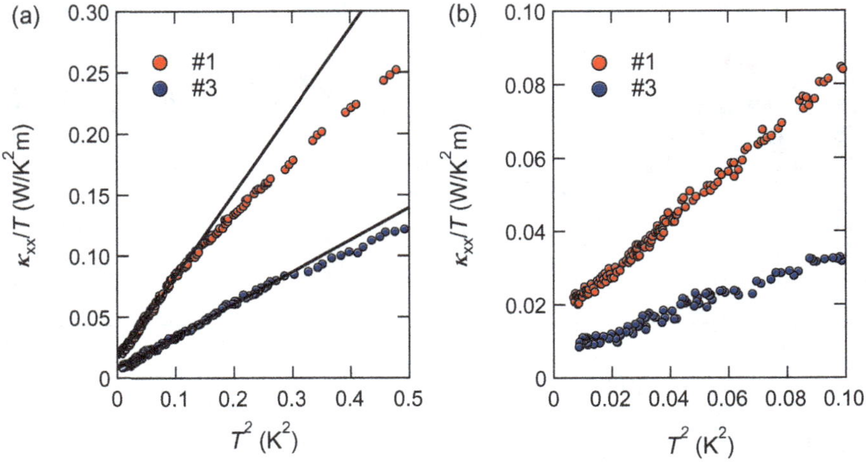

Fig. 5.12 **a** Thermal conductivity in #1 and #3 divided by temperature as a function of T^2. The solid lines represent the linear fitting performed in the low-temperature regime. **b** Expanded view of (**a**) below $T^2 = 0.1$ K. © 2021 The author.

non-zero κ^0_{xx}/T. It should also be stressed that the observed finite κ^0_{xx}/T does not originate from charged quasiparticles, in contrast to the situation in conventional metals. Evidence for this is given by the spectacular violation of the Wiedemann–Franz (WF) law, which connects the electronic thermal conductivity to the electrical resistivity (see Sect. 1.4.4 for details). In moderately pure metals at low temperatures, $L = \kappa^{\rm qp}_{xx}\rho_{xx}/T \leq L_0$ is generally satisfied, where $L_0 = \pi^2 k_{\rm B}^2/3e^2$ is the Lorenz number. The values of $\kappa^0_{xx}\rho^0_{xx}/T$ for #1 and #3 are found to be $\sim 6 \times 10^4 L_0$ and $\sim 5 \times 10^3 L_0$, respectively. It is highly unlikely that the surface metallic region significantly violates the WF law. In fact, it is well established that the WF law holds in 2D metals, even in the quantum Hall regime. The WF expectation of κ^0_{xx}/T from the metallic surface is less than 10^{-4} W/K^{2}m, which is by far smaller than the experimental resolution. These results lead us to conclude that the neutral fermions in the insulating bulk of the samples are responsible for the finite κ^0_{xx}/T. In other words, as the bulk resistivity diverges as $T \to 0$, the Lorenz number for the heat-carrying quasiparticles also diverges. Thus, the thermal conductivity and specific heat data very strongly suggest the presence of highly mobile and gapless neutral fermionic excitations. The results for YbB_{12} are in contrast to those for SmB_6, where a T-linear term is present in C [2–4] but absent in κ_{xx} at zero field [4–6].

Now, we discuss the field dependence of thermal conductivity in YbB_{12}. Figures 5.13a–b plot κ_{xx}/T as a function of T^2 in various magnetic fields for #1 and #3. κ_{xx}/T is slightly enhanced on applying a magnetic field. We fit the data using Eq. 5.21 to extract the phonon and quasiparticle contributions. The fact that all the data can be linearly fitted clearly indicates that the phonons remain in the boundary scattering regime even in the magnetic fields. The finite κ^0_{xx}/T is resolved in the whole field range up to 12 T for both samples. We also plotted κ_{xx}/T as a function of

T in Fig. 5.13c. Although the extrapolation of κ_{xx}/T to $T \to 0$ in this plot estimates the lower limit of κ_{xx}^0/T, finite intercepts are resolved in all the data, implying that the neutral fermions exist regardless of the α value (the exponent of the phonon power). Interestingly, κ_{xx}^0/T estimated by the linear fitting shows non-monotonic growth on applying fields, as shown in Fig. 5.13d. The present results imply that the neutral fermions couple to magnetic fields. Another prominent feature is that the κ_{xx}^0/T of #1 is much more enhanced by the magnetic fields than that of #3: The κ_{xx}^0/T of #1 shows a steep increase at low fields and becomes nearly twice that at 0 T, while the κ_{xx}^0/T of #3 is only slightly enhanced. Given that #1 shows a larger activation energy than #3, this result indicates that higher-quality crystals with lower impurity scattering rates exhibit a larger magneto-thermal conductivity. As larger κ_{xx}^0/T values arise from longer mean free paths, this result suggests (as might be expected) that the more mobile neutral fermions are more strongly influenced by a magnetic field.

5.3.3 *Mean Free Path of the Neutral Fermions*

The observed κ_{xx}^0/T for #1 is nearly twice that for #3, while γ for #2, the quality of which is very close to that of #1, coincides with that for #3. This quantitative difference derives from the mean free path of neutral fermions in the two types of crystals. The quasiparticle thermal conductivity is related to the specific heat by

$$\frac{\kappa_{xx}^{\mathrm{qp}}}{T} = \frac{1}{3}\gamma v_F \ell_{\mathrm{qp}} \tag{5.22}$$

where v_F is the Fermi velocity, and ℓ_{qp} is the mean free path of the neutral fermions. Therefore, the ℓ_{qp} of #1 and #2 is twice that of #3. Interestingly, this indicates that more strongly insulating crystals with larger activation energies have higher-mobility neutral quasiparticles. A fascinating question is whether the CNFs are responsible for the QOs. To examine this, we estimate ℓ_{qp} from Eq. 5.22 by assuming that v_F is given by the Fermi velocity obtained from the SdH oscillations in the insulating state, which we discussed in Sect. 3.3. By assuming a simple spherical Fermi surface, we obtain $v_F = \hbar k_F/m^* \approx 1.3 \times 10^4$ m/s from the SdH oscillations, where $k_F \approx 1.7$ nm^{-1} is the Fermi wave number, and $m^* \approx 15m_e$ is the effective mass. We estimate $\ell_{\mathrm{qp}} \approx 54$ and 25 nm, which is nearly 70 and 30 times larger than the lattice constant $a \approx 5.28$, for #1 and #3, respectively. Although the mean free path is large, the heavy effective mass leads to rather small mobilities μ:

$$\mu = \frac{e\ell_{\mathrm{qp}}}{m^* v_{\mathrm{F}}} \tag{5.23}$$

is approximately 480 cm^2/Vs (0.048 T^{-1}) for #1 and 230 cm^2/Vs (0.023 T^{-1}) for #3. According to Eq. 2.14, the reciprocal of the mobility corresponds to the character-

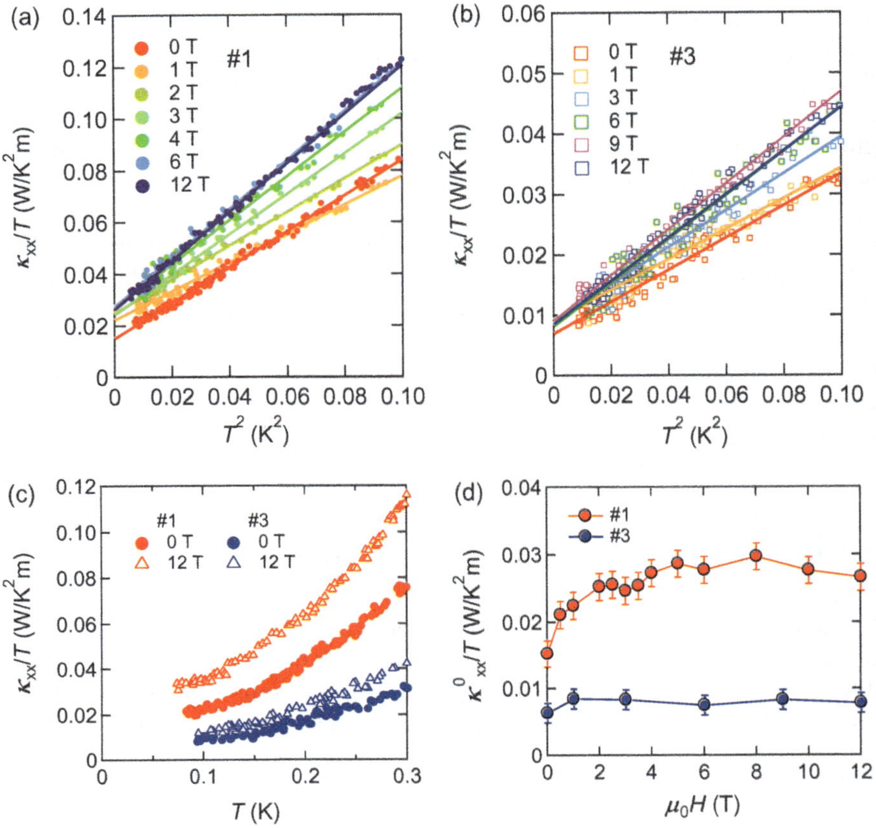

Fig. 5.13 **a**, **b** Thermal conductivity divided by temperature as a function of T^2 under magnetic fields for samples #1 and #3. The solid lines are the fitting results, showing a finite κ_{xx}^0/T throughout the field range for both samples. **c** κ_{xx}/T against T. **d** Field dependence of κ_{xx}^0/T obtained by from linear fit shown in (**a**) and (**b**). © 2021 The author.

istic magnetic fields B_c, above which QOs begin to develop. The above estimation corresponds to $B_c \sim 21$ T and 43 T for #1 and #3, respectively. This simplified model explains why magnetic fields of 30–40 T are needed to resolve the SdH oscillations in #1 and why the oscillatory amplitudes in #3 are much smaller than those in #1. Therefore, this rather crude estimation suggests that the enhanced thermal conductivity in zero field and the visible SdH effect at high fields in the sample with large ℓ_{qp} are intimately connected. The present observations for several crystals are consistent with this scenario, where CNFs are responsible for the QOs.

Finite values of both γ and κ_{xx}^0/T in the insulating state have been reported in a quantum-spin-liquid candidate, the organic compound $EtMe_3Sb[Pd(dmit)_2]_2$ (DMIT) having a 2D triangular lattice [18, 19]. In DMIT, finite γ and κ_{xx}^0/T have been discussed in terms of electrically neutral spinons forming the Fermi surface. In addition, in DMIT, a T-independent Pauli-paramagnetic-like magnetic susceptibility

χ is observed, and the Wilson ratio R_W is close to unity, which is a basic property of metals [20] (see Sect. 1.4.3 for details). As shown in Fig. 3.1c, a temperature-independent χ is also observed in YbB_{12}, and it may be ascribed to the van Vleck contribution within the J =7/2 multiplet, although this contribution has not been quantitatively estimated. An alternative explanation is that the neutral fermions give rise to a Pauli-paramagnetic-like χ. A simple estimation of R_W using the measured $\chi \approx 3 \times 10^{-3}$ emu/mol and $\gamma \approx 4$ mJ/molK^2 yields $R_W \approx 100$, which is comparable to that reported in some spin-liquid systems [21]. In DMIT, the quantum oscillations have not been observed up to 32 T, although its possible observation was suggested theoretically [22]. Thus, more experiments are needed for quantitative comparisons between these two extraordinary systems.

5.3.4 Thermal Hall Angle

As discussed in Sect. 2.6.2, an observable thermal Hall signal has been proposed within a theoretical framework, where neutral fermions are responsible for the QOs in KIs [22, 23]. The neutral fermions are expected to experience the Lorentz force as the conduction electrons in metals do, forming Landau quantization in a magnetic field and giving rise to the QOs. In an attempt to observe such an effect, we measured the thermal Hall conductivity κ_{xy}. The tangent of the thermal Hall angle

$$\tan\theta_{\mathrm{H}} = \frac{\kappa_{xy}}{\kappa_{xx}} = \omega_{\mathrm{c}}\tau \tag{5.24}$$

provides similar information as the electrical Hall angle in conventional metals. Here, $\omega_{\mathrm{c}} = (eb)_{\mathrm{eff}}/m^*$ corresponds to the cyclotron frequency of the neutral fermion, where b is the effective magnetic field experienced by neutral fermions, and τ is the scattering time. Figure 5.14a shows the field dependence of $\omega_{\mathrm{c}}^0\tau \equiv (\kappa_{xy}/T)/(\kappa_{xx}^0/T)$ at 0.2 and 0.5 K. As $\kappa_{xx} > \kappa_{xx}^0$, $\omega_{\mathrm{c}}^0\tau$ yields the upper limit for $\omega_{\mathrm{c}}\tau$. As can be observed in the data, no discernible thermal Hall effect is observed at both temperatures; $\omega_{\mathrm{c}}^0\tau$, and hence, $\omega_{\mathrm{c}}\tau$ are less than 0.005, which is less than our experimental resolution. In conventional metals with a single carrier type, $\omega_{\mathrm{c}}\tau = eB\tau/m^*$ is of the order of unity at the magnetic field where the QOs appear. As the SdH oscillations are observed around 40 T, the thermal Hall angle at 10 T could be expected to be of the order of 0.2, which is much larger than the observed thermal Hall angle. This rough estimation is also shown by the dotted line in Fig. 5.14b. As expected by the insulating behavior in resistivity, the electrical Hall angle is also vanishingly small compared to the above expectation (Fig. 5.14b), clearly implying that the conventional electrons cannot be the source of the QOs. These vanishingly small Hall angles may suggest that $(eb)_{\mathrm{eff}}$ is significantly different from eB [12, 24]. However, it is premature to conclude that the neutral fermions are not responsible for the SdH oscillations, because the small thermal Hall angle may be explained by a nonlinear B dependence of b or the presence of electron- and hole-like pockets of neutral fermions. In the latter scenario,

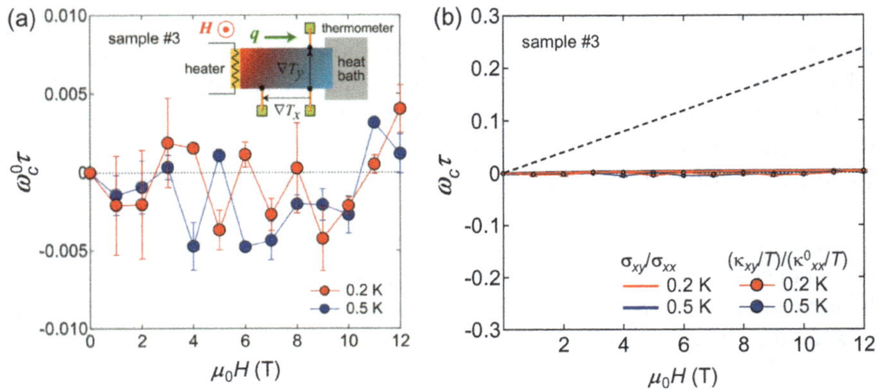

Fig. 5.14 **a**, **b** Thermal Hall angle at $T = 0.2$ and 0.5 K with different scales. The inset of (**a**) shows the setup for the thermal Hall conductivity measurements. The dotted line in (**b**) shows the B-linear Hall response expected in a conventional metal, in which QOs appear at approximately 40 T. © 2021 The author

compensation effects may reduce the thermal Hall signal considerably. The presence of a Fermi surface of neutral fermions and the coupling to an external magnetic field with a negligible thermal Hall angle call for further studies.

5.4 Conclusion

In this chapter, we presented the results of the specific heat and thermal conductivity measurements of YbB_{12} single crystals, in which we found the QOs discussed in Chap. 3, down to the dilution temperature with various magnetic fields of up to 12 T. Remarkably, terms with linear temperature dependence in both the specific heat and thermal conductivity, which are comparable to those in ordinal metals, were clearly resolved. Nevertheless, the resistivity shows clear activation-type behavior. Thus, the sizable fermionic contributions in thermal conductivity lead to the strong violation of the Wiedemann–Franz law, which reveals exotic bulk properties in YbB_{12}: It is electrically insulating but thermally metallic. Our observations, therefore, impose novel gapless excitations of CNFs. Furthermore, they are significantly affected by external fields. Although it is tempting to conclude that the observed neutral fermions are the source of the QOs at high fields, direct evidence was not provided by the thermal Hall angle measurements. Nevertheless, the scenario cannot be completely ruled out: The rough estimations of the mean free path and mobility of the neutral fermions as well as their sample dependence are consistent with an emergent Fermi surface of neutral fermions. Our findings therefore call for further theoretical and experimental investigations on KIs to explain the origin of the CNFs in YbB_{12} and their relation to the unconventional QOs observed in the insulating phase of KIs.

References

1. Matsuda Y, Izawa K, Vechter I (2006) J Phys Condens Matter 18:R705
2. Phelan WA, Koohpayeh SM, Cottingham P, Freeland JW, Leiner JC, Broholm CL, McQueen TM (2014) Phys Rev X 4:031012
3. Hartstein M, Liu H, Hsu Y-T, Tan BS, Hatnean MC, Balakrishnan G, Sebastian SE (2020) Science 23:101632
4. Hartstein M, Toews WH, Hsu YT, Zeng B, Chen X, Ciomaga Hatnean M, Zhang QR, Nakamura S, Padgett AS, Rodway-Gant G, Berk J, Kingston MK, Zhang GH, Chan MK, Yamashita S, Sakakibara T, Takano Y, Park JH, Balicas L, Harrison N, Shitsevalova N, Balakrishnan G, Lonzarich GG, Hill RW, Sutherland M, Sebastian SE (2018) Nat Phys 14:166
5. Xu Y, Cui S, Dong JK, Zhao D, Wu T, Chen XH, Sun K, Yao H, Li SY (2016) Phys Rev Lett 116:246403
6. Boulanger ME, Laliberté F, Dion M, Badoux S, Doiron-Leyraud N, Phelan WA, Koohpayeh SM, Fuhrman WT, Chamorro JR, McQueen TM, Wang XF, Nakajima Y, Metz T, Paglione J, Taillefer L (2018) Phys Rev B 97:245141
7. Fuhrman WT, Leiner J, Nikolić P, Granroth GE, Stone MB, Lumsden MD, DeBeer-Schmitt L, Alekseev PA, Mignot JM, Koohpayeh SM, Cottingham P, Phelan WA, Schoop L, McQueen TM, Broholm C (2015) Phys Rev Lett 114:036401
8. Sato Y, Xiang Z, Kasahara Y, Taniguchi T, Kasahara S, Chen L, Asaba T, Tinsman C, Murayama H, Tanaka O, Mizukami Y, Shibauchi T, Iga F, Singleton J, Li L, Matsuda Y (2019) Nat Phys 15:954
9. Wang YX, Revaz B, Erb A, Junod A (2001) Phys Rev B 63:094508
10. Taylor O (2007) Ph.D. thesis. University of Bristol
11. Chaney RC, Lafon EE, Lin CC (1971) Phys Rev B 4:2734
12. Watanabe D, Sugii K, Shimozawa M, Suzuki Y, Yajima T, Ishikawa H, Hiroi Z, Shibauchi T, Matsuda Y, Yamashita M (2016) Proc Natl Acad Sci USA 113:8653
13. Peroni I, Gottardi E, Peruzzi A, Ponti G, Ventura G (1999) Nucl Phys B - Proc Suppl 78:573
14. Czopnik A, Shitsevalova N, Pluzhnikov V, Krivchikov A, Paderno Y, Onuki Y (2005) J Phys Condens Matter 17:5971
15. Nemkovski KS, Alekseev PA, Mignot JM, Rybina AV, Iga F, Takabatake T, Shitsevalova NY, Paderno YB, Lazukov VN, Nefeodova EV, Tiden NN, Sadikov IP (2006) J Solid State Chem 179:2895
16. Grechneva GE, Baranovskiy AE, Fil VD, Ignatova TV, Kolobov IG, Logosha AV (2008) Low Temp Phys 34:1167
17. Anderson PW, Halperin BI, Varma CM (1972) Philos Mag 25:1
18. Yamashita M, Nakata N, Senshu Y, Nagata M, Yamamoto HM, Kato R, Shibauchi T, Matsuda Y (2010) Science 328:1246
19. Yamashita S, Yamamoto T, Nakazawa Y, Tamura M, Kato R (2011) Nat Commun 2:275
20. Watanabe D, Yamashita M, Tonegawa S, Oshima Y, Yamamoto HM, Kato R, Sheikin I, Behnia K, Terashima T, Uji S, Shibauchi T, Matsuda Y (2012) Nat Commun 3:1090
21. Balents L (2010) Nature 464:199
22. Katsura H, Nagaosa N, Lee PA (2010) Phys Rev Lett 104:066403
23. Chowdhury D, Sodemann I, Senthil T (2018) Nat Commun 9:1766
24. Motrunich OI (2006) Phys Rev B 73:155115

Chapter 6
Conclusion

The textbook definition of metals is that they are materials which possessing a Fermi surface in momentum space. The presence of a Fermi surface is manifested by observations of quantum oscillations (QOs) in magnetic fields and T-linear terms in specific heat and thermal conductivity. We have presented a series of experiments on the Kondo insulator (KI) YbB_{12} and revealed exotic electronic properties.

We reviewed that YbB_{12} shows QOs not only in magnetization, but also in resistivity (Shubnikov–de Haas (SdH) effect), although it is electrically insulating [1, 2]. The SdH oscillations still occur above the insulator-metal transition critical field, and the oscillatory amplitude shows Lifshitz–Kosevich behavior in both states, indicating that fermions contribute to Landau quantization. Angle-resolved experiments captured the 3D geometry of the frequency, suggesting that the insulating bulk possesses an unconventional Fermi surface of charge neutral fermions (CNFs). Interestingly, the results in the Kondo metallic state in higher magnetic fields indicate that the Fermi surface of CNFs still contributes to the SdH effect, although it progressively depopulates with increasing field [3]. CNFs also exhibit intriguing properties such as an extremely small g-factor and the field enhancement of the cyclotron mass m^*. Moreover, the large Sommerfeld coefficient γ and FL transport coefficient A_2 in the field-induced metallic state show strong deviation from the universal Kadowaki–Woods ratio. The results indicate the coexistence of two fluids in the metallic state, CNFs, and the conventional charged heavy fermions. They strongly scatter each other and may give rise to the non-FL transport. The pulsed-field experiments revealed highly exotic electronic properties of KIs.

Topological surface conduction in YbB_{12} is investigated [4] in Chap. 4. Our transport experiments in a series of single crystals and microstructures fabricated by focused ion beam method reveal resistivity plateau at low temperatures, the amplitude of which is linearly increases against sample thickness. Moreover, we observed that the magnetoresistance shows weak-antilocalization effect at low fields, indicating that YbB_{12} is a promising candidate of topological Kondo insulator. Although

Y. Sato, *Quantum Oscillations and Charge-Neutral Fermions in Topological Kondo Insulator YbB_{12}*, Springer Theses,
https://doi.org/10.1007/978-981-16-5677-4_6

the metallic surface state dominates its electric transport at low temperatures, we found the microstructure does not exhibit SdH oscillations at high fields. Our results provide a supporting evidence that the SdH effect derives from the insulating bulk, rather than topological surface states, which is consistent with the discussions in Chap. 3.

We also conducted low-temperature specific heat and thermal conductivity measurements with various fields to study low-energy excitations in YbB_{12} [2], which is presented in Chap. 5. Surprisingly, we observed finite T-linear terms in both specific heat and thermal conductivity, which are comparable to those of ordinal metals. The sizable fermionic thermal conductivity leads to a remarkable violation of the WF law, indicating that YbB_{12} is electrically insulating but thermally metallic. The present results suggest that charge-neutral gapless fermionic excitations occur in the ground state of YbB_{12} and are highly mobile. Moreover, despite the charge neutrality, they are influenced by applied magnetic fields. Although it is tempting to link the observed fermionic thermal transport to the charge-neutral Fermi surface detected in the pulsed-field experiments, our thermal Hall conductivity measurements did not provide direct evidence to link these observations. Thus, we leave the problem of the relationship between the mobile neutral fermions and the QOs for future work.

Thus far, we have discussed a series of results providing evidence for "metallic" behaviors in the KI YbB_{12}, although it's bulk is an excellent charge insulator. Our studies revealed topological SS and highly unconventional emergent quasiparticles in this exotic quantum state, which are expected to motivate further investigations in the future.

References

1. Xiang Z, Kasahara Y, Asaba T, Lawson B, Tinsman C, Chen L, Sugimoto K, Kawaguchi S, Sato Y, Li G, Yao S, Chen YL, Iga F, Singleton J, Matsuda Y, Li L (2018) Science 69:65
2. Sato Y, Xiang Z, Kasahara Y, Taniguchi T, Kasahara S, Chen L, Asaba T, Tinsman C, Murayama H, Tanaka O, Mizukami Y, Shibauchi T, Iga F, Singleton J, Li L, Matsuda Y (2019) Nat Phys 15:954
3. Xiang Z, Chen L, Chen KW, Tinsman C, Sato Y, Asaba T, Lu H, Kasahara Y, Jaime M, Balakirev F, Iga F, Matsuda Y, Singleton J, Li L (2021) Nat Phys 17:788
4. Sato Y, Xiang Z, Kasahara Y, Kasahara S, Chen L, Tinsman C, Iga F, Singleton J, Nair NL, Maksimovic N, Analytis JG, Li L, Matsuda Y (2021) J Phys D Appl Phys 54:404002

The manufacturer's authorised representative in the EU is Springer Nature Customer Service Centre GmbH, Europaplatz 3, 69115 Heidelberg, Germany. If you have any concerns regarding our products, please contact ProductSafety@springernature.com

Printed and bound by CPI Group (UK) Ltd, Croydon, CR0 4YY
30/09/2026
02225306-0001